A DELTA RENEWED
A Guide to Science-Based Ecological Restoration in the Delta

A Report of the Delta Landscapes Project:
Management Tools for Landscape-Scale Restoration of Ecological Functions

PREPARED FOR THE CALIFORNIA DEPARTMENT OF FISH AND WILDLIFE
AND THE ECOSYSTEM RESTORATION PROGRAM

NOVEMBER 2016

PROJECT DIRECTION
Julie Beagle
Letitia Grenier
Robin Grossinger

PRIMARY AUTHORS
April Robinson
Samuel Safran
Julie Beagle

CONTRIBUTING AUTHORS
Letitia Grenier
Robin Grossinger
Erica Spotswood
Scott Dusterhoff
Amy Richey

DESIGN AND LAYOUT
Ruth Askevold
Samuel Safran

SFEI
A·S·C

PREPARED BY San Francisco Estuary Institute-Aquatic Science Center

IN COOPERATION WITH AND FUNDED BY California Department of Fish and Wildlife
Ecosystem Restoration Program

CALIFORNIA
DEPARTMENT OF
FISH &
WILDLIFE

ECOSYSTEM
RESTORATION
PROGRAM
California Department of Fish & Wildlife
National Oceanic & Atmospheric Administration
US Fish & Wildlife Service

SFEI-ASC
PUBLICATION #799

THE SACRAMENTO-SAN JOAQUIN DELTA
modern waterways, islands, and tracts

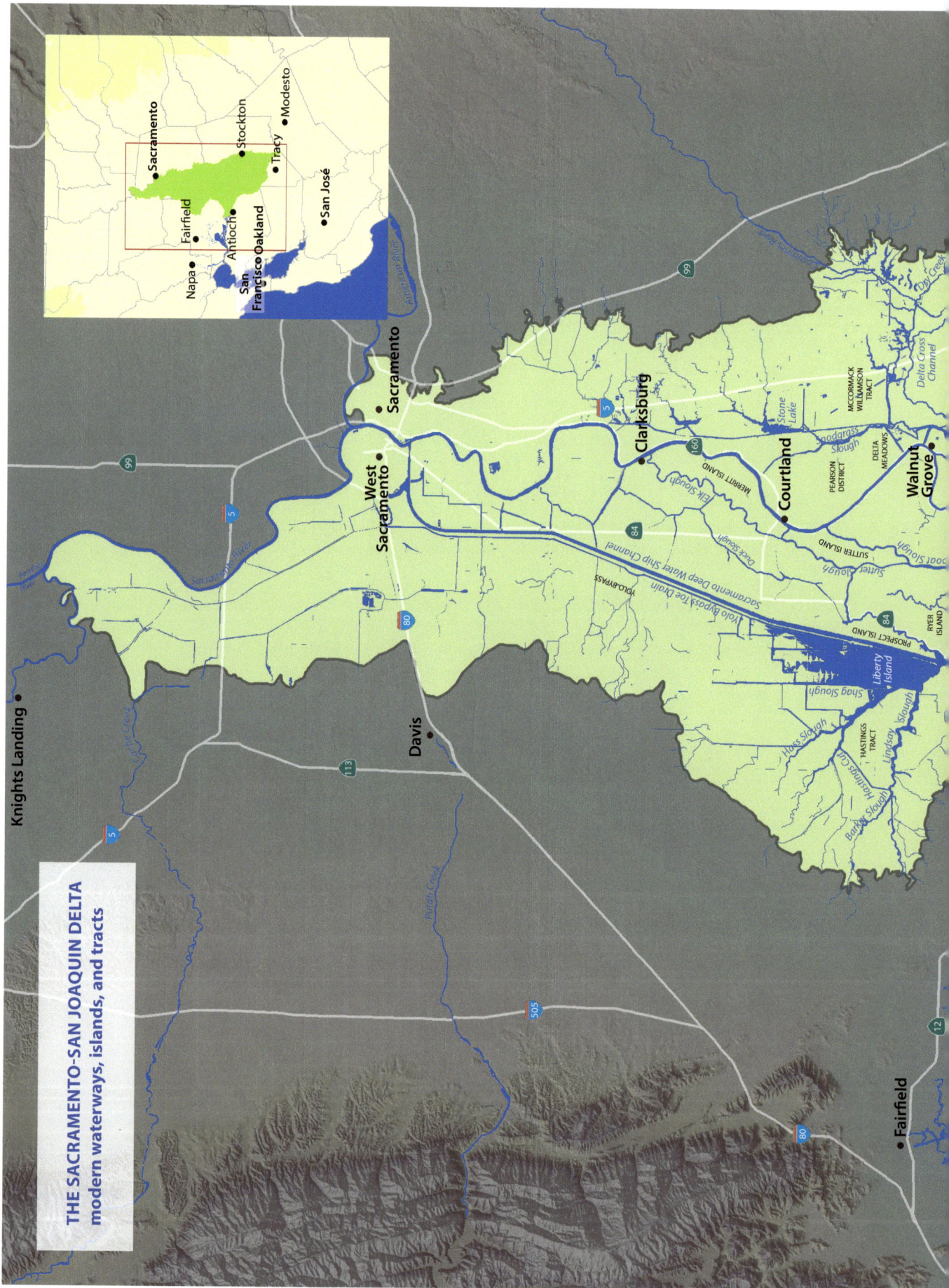

Knights Landing

Davis

West Sacramento

Sacramento

Clarksburg

Courtland

Walnut Grove

MERRITT ISLAND

Elk Slough

Sutter Slough

SUTTER ISLAND

Duck Slough

Sacramento Deep Water Ship Channel

YOLO BYPASS

Yolo Bypass Toe Drain

PROSPECT ISLAND

RYER ISLAND

Liberty Island

Shag Slough

Lindsay Slough

HASTINGS TRACT

Hastings Cut

Barker Slough

Hass Slough

Stone Lake

McCORMACK WILLIAMSON TRACT

Snodgrass Slough

PEARSON DISTRICT

DELTA MEADOWS

Delta Cross Channel

Dry Creek

Sacramento River

Putah Creek

Cache Creek

American River

Fairfield

Napa

San Francisco

Oakland

Antioch

Tracy

Stockton

Sacramento

Fairfield

Modesto

San José

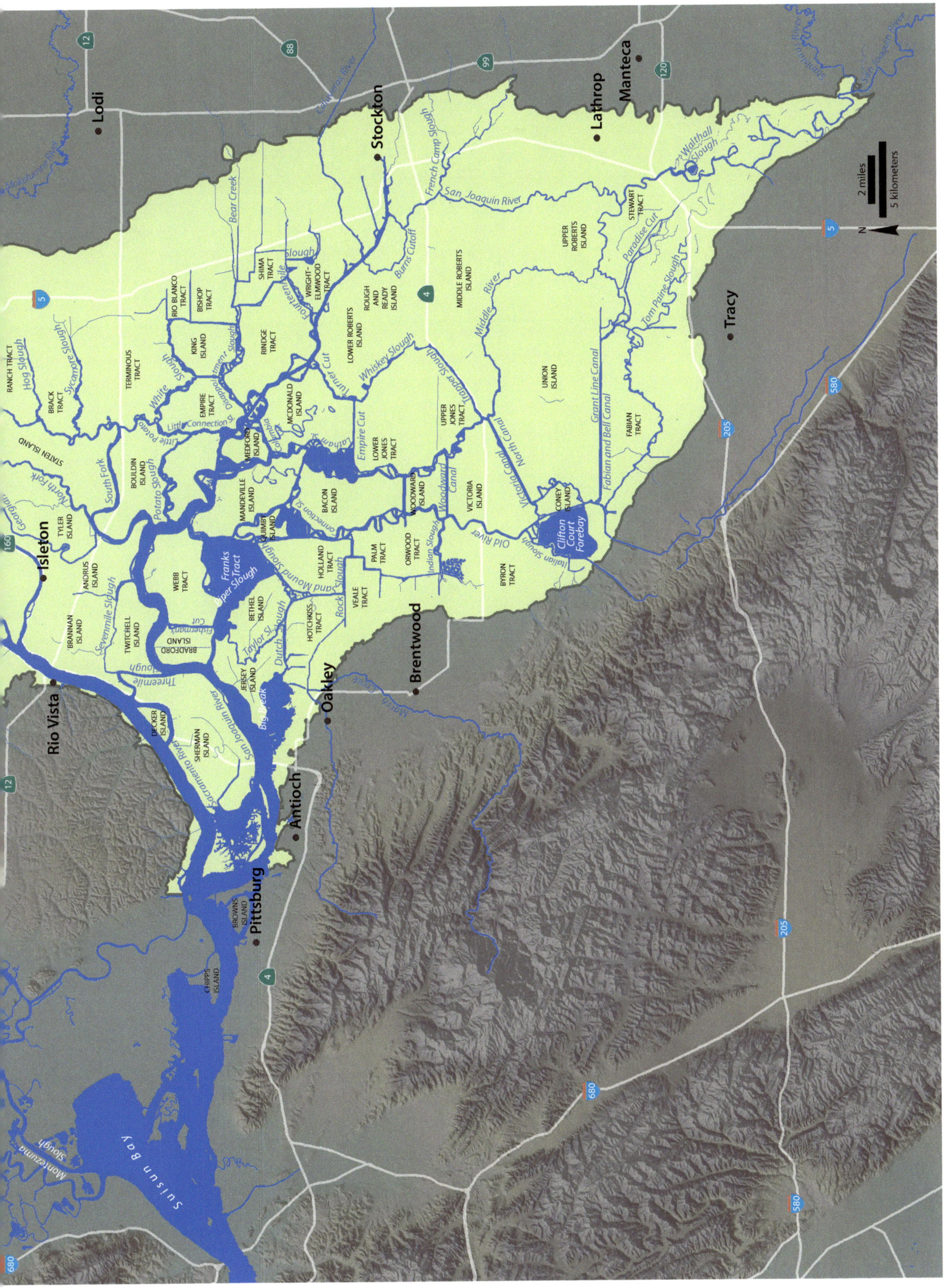

Lodi
Stockton
Lathrop
Manteca
Tracy
Isleton
Rio Vista
Brentwood
Oakley
Antioch
Pittsburg
Suisun Bay
Montezuma Slough
Franks Tract
Clifton Court Forebay
RANCH TRACT
Hog Slough
Sycamore Slough
BRACK TRACT
TERMINOUS TRACT
STATEN ISLAND
North Fork
South Fork
TYLER ISLAND
ANDRUS ISLAND
BRANNAN ISLAND
TWITCHELL ISLAND
BRADFORD ISLAND
Sevenmile Slough
Threemile Slough
DECKER ISLAND
SHERMAN ISLAND
Sacramento River
San Joaquin River
JERSEY ISLAND
Big Break
BETHEL ISLAND
Taylor Slough
Dutch Slough
HOTCHKISS TRACT
WEBB TRACT
Piper Slough
Fishermans Cut
BOULDIN ISLAND
Potato Slough
Little Potato Slough
White Slough
EMPIRE TRACT
Little Connection Sl.
KING ISLAND
RIO BLANCO TRACT
BISHOP TRACT
RINDGE TRACT
SHIMA TRACT
WRIGHT-ELMWOOD TRACT
Fourteenmile Slough
Disappointment Slough
Bear Creek
MEDFORD ISLAND
MCDONALD ISLAND
MANDEVILLE ISLAND
QUIMBY ISLAND
HOLLAND TRACT
Sand Mound Slough
Rock Slough
VEALE TRACT
PALM TRACT
Connection Sl.
BACON ISLAND
Latham Sl.
Empire Cut
LOWER JONES TRACT
UPPER JONES TRACT
WOODWARD ISLAND
Woodward Canal
ORWOOD TRACT
Indian Slough
Old River
VICTORIA ISLAND
BYRON TRACT
Italian Slough
North Canal
Victoria Canal
CONEY ISLAND
UNION ISLAND
Grant Line Canal
Fabian and Bell Canal
FABIAN TRACT
Middle River
MIDDLE ROBERTS ISLAND
LOWER ROBERTS ISLAND
ROUGH AND READY ISLAND
Whiskey Slough
Trapper Slough
Turner Cut
Burns Cutoff
French Camp Slough
UPPER ROBERTS ISLAND
STEWART TRACT
Paradise Cut
Tom Paine Slough
Walthall Slough
BROWNS ISLAND
CHIPPS ISLAND
2 miles
5 kilometers
N

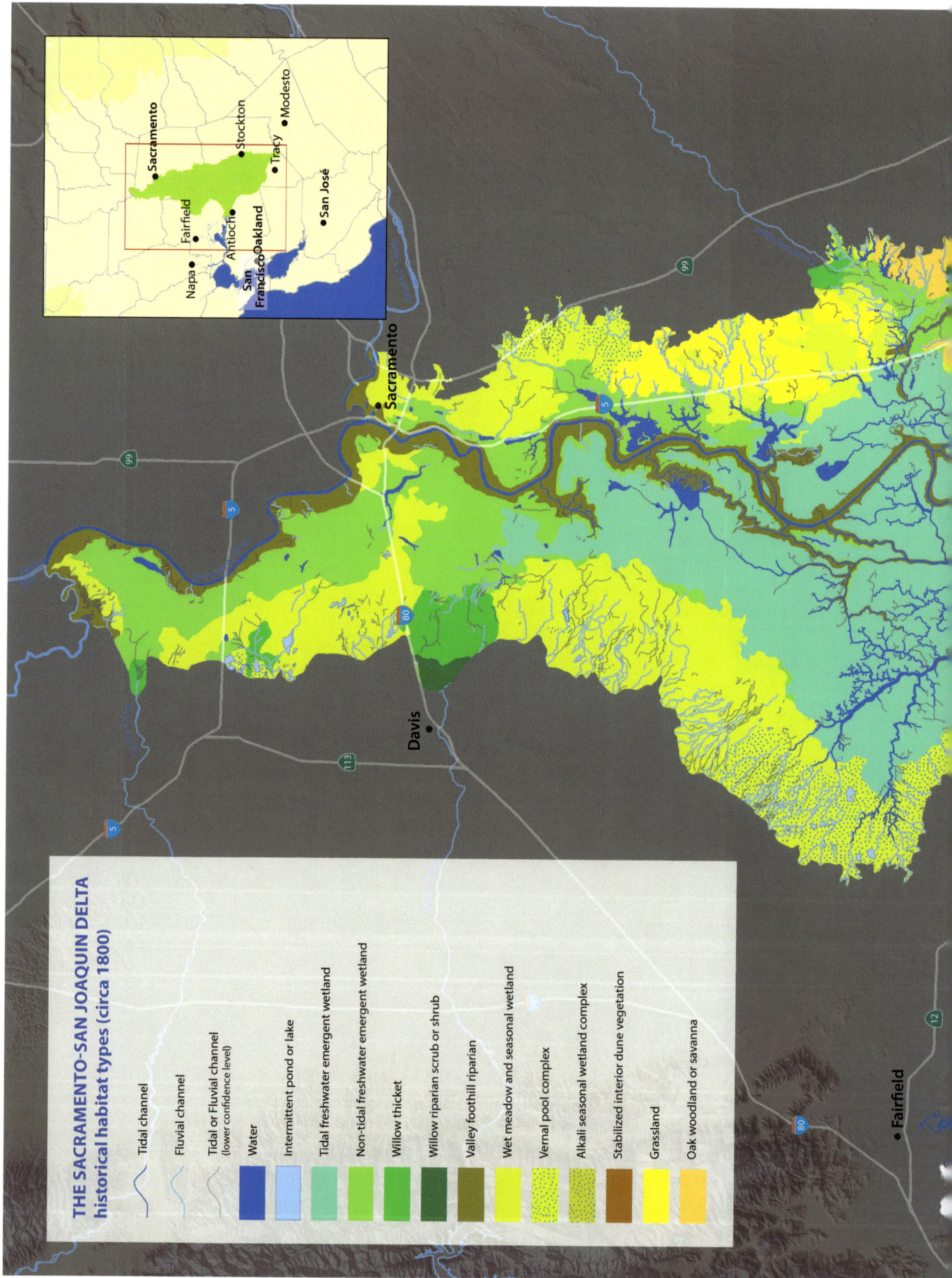

THE SACRAMENTO-SAN JOAQUIN DELTA
historical habitat types (circa 1800)

Tidal channel

Fluvial channel

Tidal or Fluvial channel
(lower confidence level)

Water

Intermittent pond or lake

Tidal freshwater emergent wetland

Non-tidal freshwater emergent wetland

Willow thicket

Willow riparian scrub or shrub

Valley foothill riparian

Wet meadow and seasonal wetland

Vernal pool complex

Alkali seasonal wetland complex

Stabilized interior dune vegetation

Grassland

Oak woodland or savanna

Sacramento

Stockton

Modesto

Tracy

San José

Fairfield

Antioch

Oakland

Napa

San
Francisco

Davis

iv

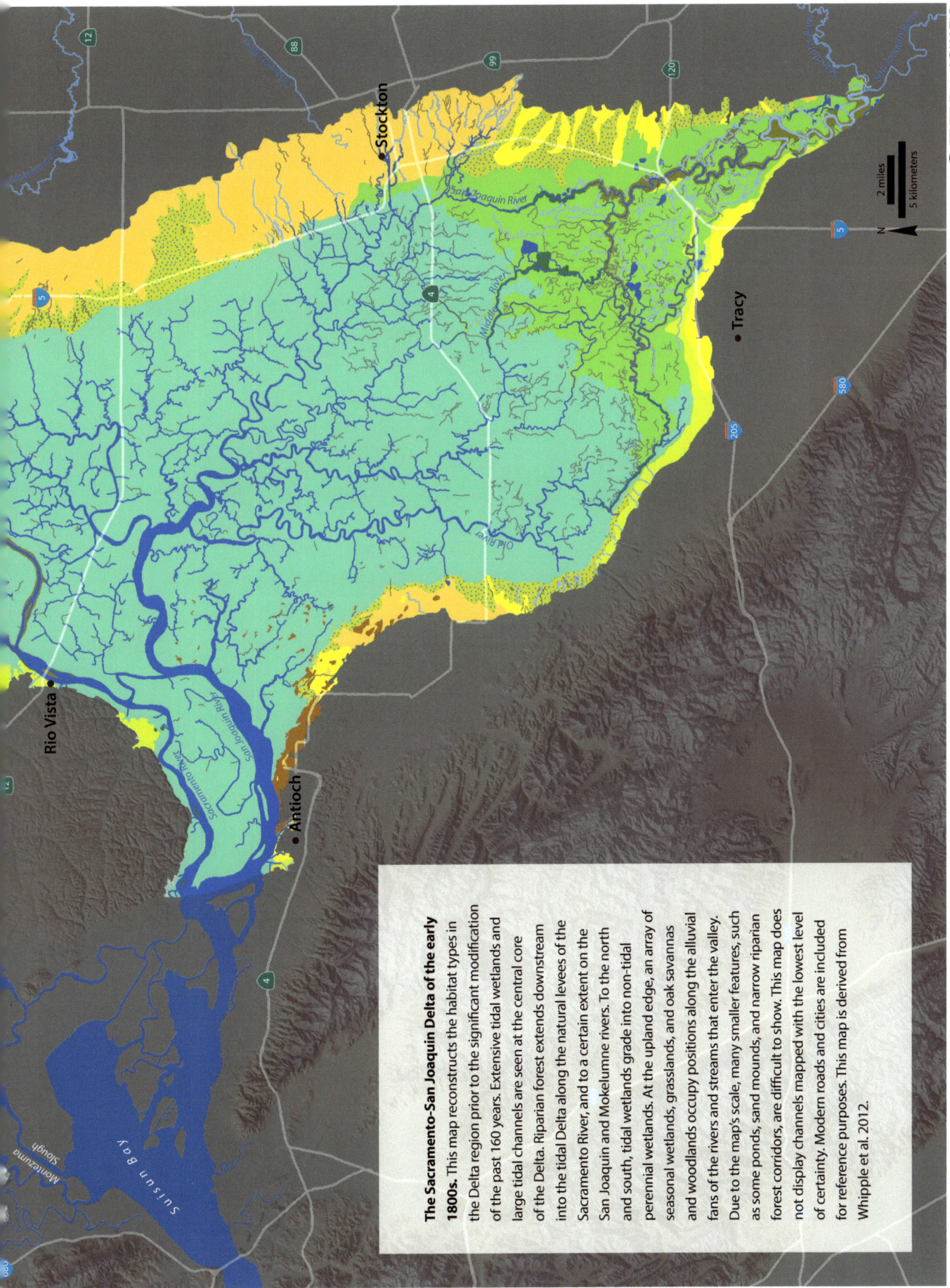

The Sacramento-San Joaquin Delta of the early 1800s. This map reconstructs the habitat types in the Delta region prior to the significant modification of the past 160 years. Extensive tidal wetlands and large tidal channels are seen at the central core of the Delta. Riparian forest extends downstream into the tidal Delta along the natural levees of the Sacramento River, and to a certain extent on the San Joaquin and Mokelumne rivers. To the north and south, tidal wetlands grade into non-tidal perennial wetlands. At the upland edge, an array of seasonal wetlands, grasslands, and oak savannas and woodlands occupy positions along the alluvial fans of the rivers and streams that enter the valley. Due to the map's scale, many smaller features, such as some ponds, sand mounds, and narrow riparian forest corridors, are difficult to show. This map does not display channels mapped with the lowest level of certainty. Modern roads and cities are included for reference purposes. This map is derived from Whipple et al. 2012.

Stockton

Tracy

Rio Vista

Antioch

San Joaquin River

Middle River

Old River

Sacramento River

San Joaquin River

Suisun Bay

Montezuma Slough

N

2 miles
5 kilometers

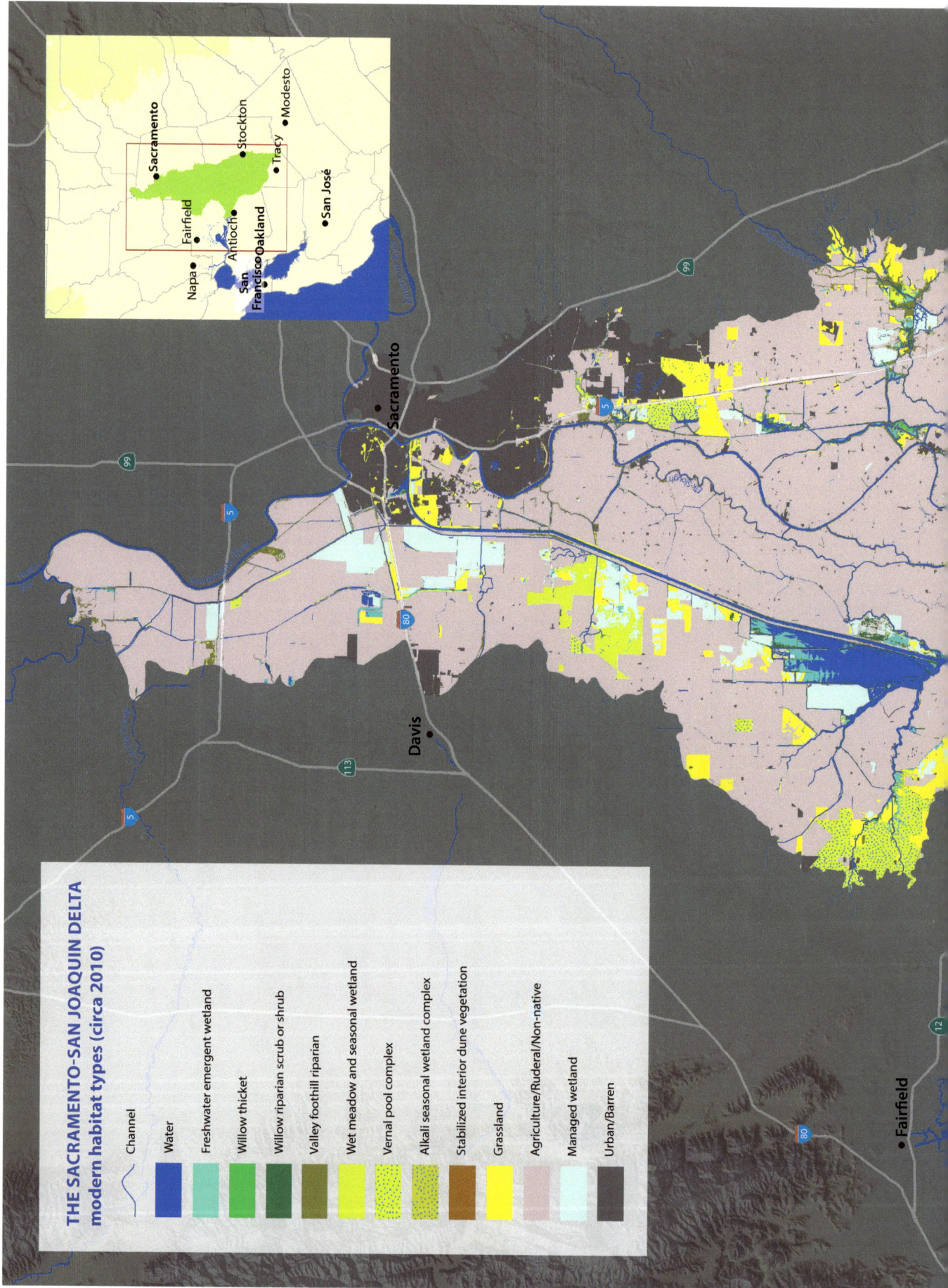

THE SACRAMENTO-SAN JOAQUIN DELTA
modern habitat types (circa 2010)

- Channel
- Water
- Freshwater emergent wetland
- Willow thicket
- Willow riparian scrub or shrub
- Valley foothill riparian
- Wet meadow and seasonal wetland
- Vernal pool complex
- Alkali seasonal wetland complex
- Stabilized interior dune vegetation
- Grassland
- Agriculture/Ruderal/Non-native
- Managed wetland
- Urban/Barren

Sacramento

Stockton

Tracy

Modesto

San José

Fairfield

Antioch

Oakland

Napa

San Francisco

Davis

Stockton

Tracy

Rio Vista

Antioch

San Joaquin River

Joaquin River

Middle River

Old River

Suisun Bay

Montezuma Slough

2 miles

5 kilometers

N

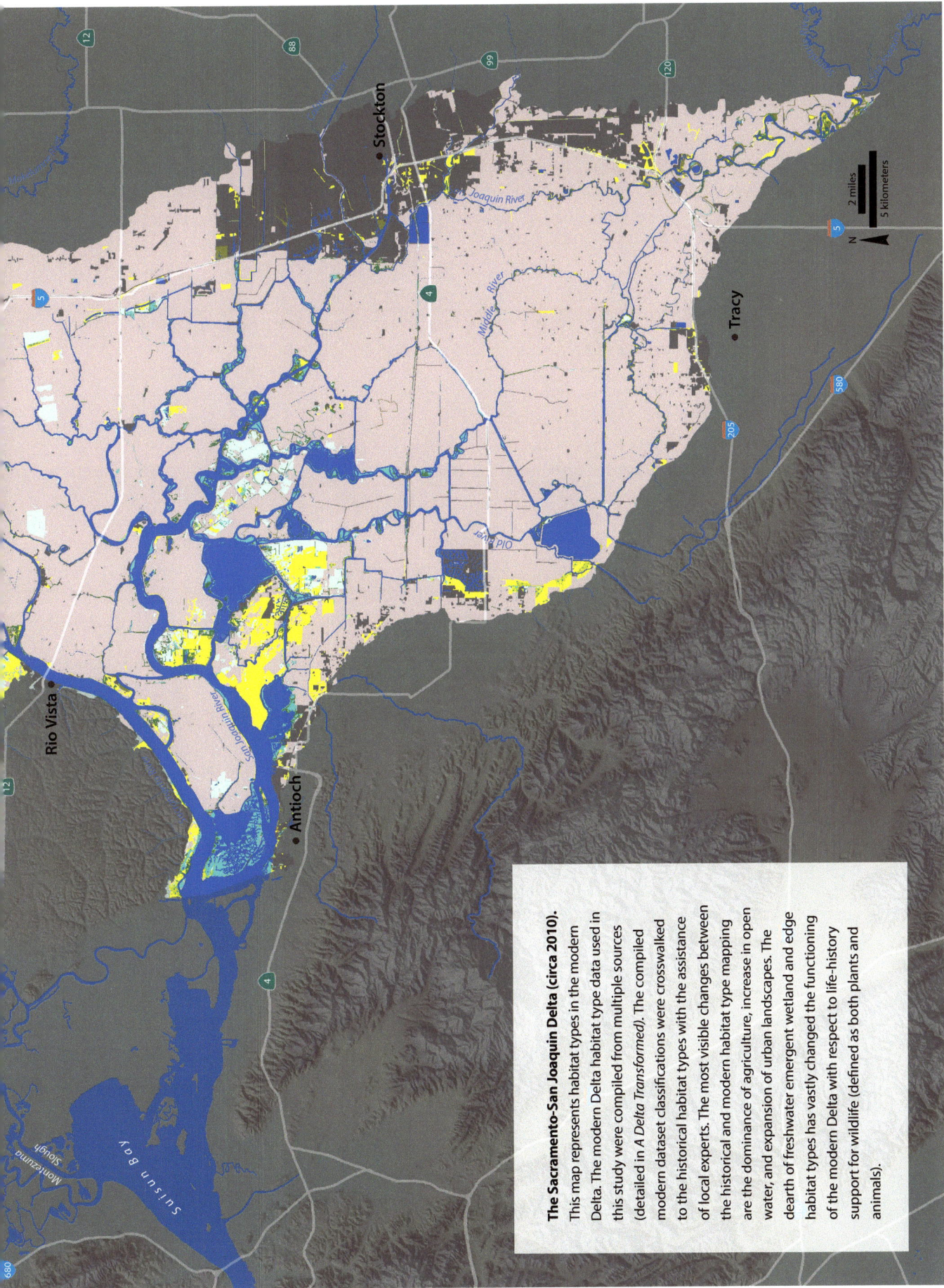

The Sacramento-San Joaquin Delta (circa 2010).
This map represents habitat types in the modern
Delta. The modern Delta habitat type data used in
this study were compiled from multiple sources
(detailed in *A Delta Transformed*). The compiled
modern dataset classifications were crosswalked
to the historical habitat types with the assistance
of local experts. The most visible changes between
the historical and modern habitat type mapping
are the dominance of agriculture, increase in open
water, and expansion of urban landscapes. The
dearth of freshwater emergent wetland and edge
habitat types has vastly changed the functioning
of the modern Delta with respect to life-history
support for wildlife (defined as both plants and
animals).

October 2016

The Delta Reform Act identifies the coequal goals of improving statewide water supply reliability and protecting, restoring, and enhancing the Delta ecosystem. This report is intended to support implementation of the goal of restoring the Delta ecosystem. It is the culmination of work begun with the *Delta Historical Ecology Investigation* to better understand the historical Delta and to serve as a guide for improving the ecological integrity of the Delta of today. The Delta can not be returned to its historical condition. The intent is to use our understanding of its historical form and function to guide future efforts to restore and enhance habitat as projects are planned and implemented in a way that supports the attributes of the historical Delta in a landscape context.

A Delta Renewed builds on previous reports examining the historical Delta and how it has been transformed into its current condition as a unique rural, agricultural oriented landscape that retains many ecologically important components. As described in the Delta Reform Act, the coequal goals are to be achieved in a manner that protects and enhances the unique cultural, recreational, natural resource, and agricultural values of the Delta as an evolving place.

Development of the report relied on the input of experts from a variety of technical backgrounds in species biology, ecological processes and physical processes. These experts provided guidance and recommendations in the interpretation of the historical form and function of the Delta and how that relates to today's Delta and future ecosystem enhancement and restoration.

While the report and its recommendations have been developed with a high level of technical input, it has been written for a general audience. The authors of the report have prepared it with the intent that the citizens of the Delta and those involved in management and decision making have a common understanding of the habitat changes needed for improved ecological health. Implementation will require the broad community of the Delta working together collaboratively. While the report identifies potential opportunities for enhancement and restoration from an ecological and physical perspective, these are possibilities and will require engagement with affected communities and stakeholders to be realized. Any actions suggested in the report are meant to be implemented voluntarily.

Restoring the Delta ecosystem will occur over long timeframes in an ever changing social, ecological, and regulatory environment affected dramatically by climate change and associated sea level rise. It will need to be done through an adaptive management process with clear objectives and performance measures, and constant learning and assessment of what has been done. This report provides the historical and landscape perspective that is a critically important foundation to our ability to successfully improve the ecological integrity of the Delta.

—**Carl Wilcox**
Policy Advisor to the Director for the Delta,
CA Dept. of Fish and Wildlife

—**Campbell Ingram**
Executive Officer, *Delta Conservancy*

—**Jessica Pearson**
Executive Officer, *Delta Stewardship Council*

October 2016

Management Tools for Landscape-Scale Restoration of Ecological Functions

The Sacramento-San Joaquin Delta is an extremely complex landscape system that supports extensive cities, varied agriculture, recreational opportunities, commercial fisheries, and a huge water management apparatus, delivering water statewide for human enterprises. It also contains remnants of a vast system of marshes and waterways that historically supported a large and highly diverse endemic biota.

Multiple pressures from the growth in the human population of California are continuing to change the Delta ecosystem, leading to a decline in its ability to support the diversity of human benefits that it once provided. It is clear that we must proactively address this dilemma and devise ways to rehabilitate populations of native species and essential ecological functions of this remarkable landscape. If this is not addressed now, we confidently predict that the Delta will continue to change to a much less desirable state.

The Delta was altered over many decades; thus, reversing the decline of native plants and animals has proven difficult. Recovering native species that are adapted to historical habitat types and processes will require a new approach that emphasizes landscape repair processes at appropriate scales in space and time. These investments require large spaces that are bounded by natural features—channels, wetlands and uplands—not levees and roads. We worry that reconnecting only small properties on subsided land bounded by levees will produce uncertain recovery regimes, limited ecosystem value, and social problems. A long view of landscape rehabilitation also affords flexibility for land acquisition, management of tidal energy, integration with human uses, and interim co-benefits from carbon sequestration, recreation, and terrestrial species conservation.

At this time the Delta is profoundly changed, perhaps irreversibly, by modified flow patterns, degraded water quality, alien species, land subsidence, and simplified, over-connected waterways. Robust policies that address some of these problems (flows, quality, alien species) could improve conditions quickly. However, reconciling competing values may render those measures only marginally effective. Recovering landscape forms, habitat types, and processes that favor native species is a more complex and longer-term undertaking. Land acquisition for restoration and environmental permitting are particularly challenging. Political pressures and policy timelines to complete projects quickly often run counter to the culture of risk-averse permitting agencies.

A Delta Renewed and the larger *Delta Landscapes* project are part of an ongoing effort to address these critically important challenges. They provide guidance for restoration that renews critical functions over time without having major negative impacts on the current human enterprise in the Delta area and on other areas dependent on its services. The idea is to determine the most efficient ways to recover Delta functioning so that human benefits are maintained, while investing in process-based recovery of landscape functions. The scientists

(CONTINUED NEXT PAGE)

working on this project assume that this goal is best achieved if we can understand how the Delta functioned before major European modification. We can then use that knowledge as a guide to developing a diverse portfolio of habitats and connections that might sustain native species, and the attributes of a healthy Delta into the future.

The effort to reconstruct the Delta of the early 1800s has been challenging, but is now completed. The next step is to devise management scenarios that will recover important functions and native species where possible. There is no intent to try to recreate the Delta of the past. That is impossible without sacrificing our objective of maintaining and even improving the human enterprises now present. What is essential now is to obtain general agreement among all of the stakeholders, which includes all citizens of California, that the Delta should be managed for the benefit of people and wildlife in a sustainable way. Given the uncertainties inherent in any large-scale reconciliation effort, there is a real need to appreciate the complex interactions and inter-dependencies that characterize the various components of the Delta landscape. There is also a need to understand the uncertainties of future climate change and to recognize the diversity of needs among our fellow citizens. Our hopes for success will ultimately depend on generating a spirit of cooperation for the common good. It won't be easy, but the stakes are high. The *Delta Landscapes* project is dedicated to that goal.

Stephanie Carlson, *University of California, Berkeley*
James Cloern, *U.S. Geological Survey*
Christopher Enright, *Delta Science Program*
Geoffrey Geupel, *Point Blue Conservation Science*
Todd Keeler-Wolf, *California Department of Fish and Wildlife*
William Lidicker, Jr., *University of California, Berkeley*
Steve Lindley, *National Marine Fisheries Service (NOAA)*
Jay Lund, *University of California, Davis*
Peter Moyle, *University of California, Davis*
Anke Mueller-Solger, *U.S. Geological Survey*
Hildie Spautz, *California Department of Fish and Wildlife*
Alison Whipple, *University of California, Davis*

— OF THE **Landscape Interpretation Team Science Advisors**

* Please note that this preface and the recommendations in this report reflect the technical expertise of the LIT but do not necessarily reflect the positions of the agencies they work for.

Acknowledgments

This project was funded by the California Department of Fish and Wildlife (CDFW) through the Ecosystem Restoration Program (ERP). We give special thanks to Carl Wilcox of CDFW and Cliff Dahm (Delta Science Program [DSP] lead scientist), both of whom helped shape the project. We would like to thank Daniel Burmester and Kevin Fleming, our CDFW project managers, for their technical advice and support.

The project has benefited substantially from the sound guidance, engagement, and enthusiasm of our technical advisory group, the *Landscape Interpretation Team:* Stephanie Carlson (University of California, Berkeley [UCB]), Jim Cloern (United States Geological Survey [USGS]), Brian Collins (University of Washington) Chris Enright (DSP), Joe Fleskes (USGS), Geoff Geupel (Point Blue), Todd Keeler-Wolf (CDFW), William Lidicker (UCB), Steve Lindley (National Oceanic and Atmospheric Administration, National Marine Fisheries Service), Peter Moyle (University of California, Davis [UCD]), Anke Mueller-Solger (USGS), Hildie Spautz (CDFW), Alison Whipple (UCD) and Dave Zezulak (CDFW). Other key advisors to whom we are indebted include Brian Atwater (University of Washington), Jay Lund (UCD), and John Wiens (Colorado State University).

We are indebted to Alison Whipple (UCD) who, in addition to conceiving of this project, has provided valuable insight and technical assistance along the way as part of the Landscape Interpretation Team.

We give thanks to our separate team of advisors dedicated to discussing functions relating to support for fish populations including: Carson Jeffres (UCD), Ted Sommer (California Department of Water Resources [DWR]), John Durand (UCD), and Jim Hobbs (UCD). We appreciate the advice gained from meetings with Rodd Kelsey and Jennifer Beringer (The Nature Conservancy [TNC]) regarding waterbird use of flooded agricultural lands in the Delta. We are also grateful for input from Greg Yarris (U.S. Fish and Wildlife Service) regarding waterbirds and agriculture, and Maya Kepner (American West Conservation) regarding wildlife-friendly agriculture.

We could not have completed this project without the invaluable support, review, and coordination of staff members of several Delta agencies, including Scott Cantrell, Brooke Jacobs, and Christina Sloop (CDFW); Darcy Austin, Marina Brand, Lindsay Correa, Lauren Hastings, and Rainer Hoenicke (DSP); Jessica Davenport and Jessica Law (Delta Stewardship Council); and Campbell Ingram (Delta Conservancy).

We are grateful to Yiping Lu and Rafael Tiffany from UC Berkeley for their work rendering images for this report.

Finally, we are grateful to many SFEI-ASC staff members who contributed to this project: Shira Bezalel, Sean Baumgarten, Warner Chabot, Emily Clark, Kate Roberts (intern), and Micha Salomon.

SUMMARY

This report offers guidance for creating and maintaining landscapes in the Sacramento-San Joaquin Delta that support desired ecological functions, while retaining the overall agricultural character and water-supply service of the region. Based on extensive research into how the Delta functioned historically, how it has changed, and how it is likely to evolve, we discuss where and how to re-establish the dynamic natural processes that can sustain native Delta habitats and wildlife into the future. The approach, building on work others have piloted and championed, is to restore or emulate natural processes where possible, establish an appropriate mosaic of habitat types at the landscape scale, use multi-benefit management strategies to increase support for native species in agricultural and urban areas, and allow the Delta to adapt to future uncertainties of climate change, levee failure, and human population growth. With this approach, it will be critical to integrate ecological improvements with the human landscape: a robust agricultural economy, water infrastructure and diversions, and urbanized areas. Strategic restoration that builds on the history and ecology of the region can contribute to the strong sense of place and recreational value of the Delta.

THE CHALLENGE OF COMPLEXITY

There are no easy solutions for managing such a complex system that serves so many purposes. The Delta supplies freshwater to a large portion of California's cities and agriculture, supports an agricultural economy and culture, and is home to native wildlife found nowhere else in the world. Although agencies, stakeholders, scientists, and planners have attempted to coalesce around a vision for the future Delta for many years, the region remains hampered by many challenges, including poor water quality, an over-allocated water supply, decaying infrastructure, invasions of alien species, novel ecosystems that no longer support desired functions, and a complex management structure.

OUR APPROACH

A Delta Renewed attempts to inform and contribute to ongoing planning efforts by providing a science-based, big-picture perspective on how to re-establish a landscape that functions well for people and native wildlife. We offer regional recommendations and on-the-ground strategies to help contextualize, design, implement, and manage future Delta landscapes that can support desired ecological functions over the long term, like healthy native fish populations, a productive food web, and support for endangered species. To develop this guidance, we evaluated the landscape patterns and processes that supported wildlife in the historical Delta, measured how they have changed, and assessed the potential for establishing smaller, modified landscapes in the future Delta that are resilient, productive, sustainable, and supportive of people and native wildlife. The report contributes to Delta planning by providing a large-scale, long-time-frame perspective on restoration opportunities, using a systems approach designed to benefit a holistic suite of desired ecological functions, not just a few rare species. A Delta Renewed is a blueprint for creating new, reconciled landscapes that integrate natural and cultural processes, and maximize resilience to climate change, invasive species, and other challenges.

PROCESS-BASED STRATEGIES

Restoration and management actions that incorporate naturalistic physical processes are essential to the future of the Delta, particularly in light of sea-level rise and other future changes. The long-term aim of process-based restoration is to create dynamic, resilient ecosystems, rather than static habitat patches and rigid engineered structures. Restoration of critical processes, such as beneficial flooding and sediment delivery, creates and maintains habitats, fuels the food web, and enables ecosystems to recover after disturbance and continue to support native wildlife as baselines shift, which will become more important as climate change accelerates. This approach requires large spatial scales, long time frames, and coordination of complex management regimes across the Delta landscape, including a multitude of landowners, regulations, and land-uses.

The critical processes to restore are organized into five major zones in the Delta. We detail strategies for restoring:

- *tidal zone processes* in intertidal areas, channels and flooded islands, and subsided areas;

- *tidal-fluvial transition zone processes* by improving the connection between streams and floodplains;

- *fluvial processes* along streams and their floodplains as they enter the Delta;

- *wetland-terrestrial transition zone and terrestrial processes* around the periphery of the Delta; and

- *ecological processes within areas of human land use,* through wildlife-friendly farming and urban greening.

These strategies fit into the current and future Delta landscapes in ways that may not duplicate historical locations and configurations. Supporting native wildlife and other ecological functions in the Delta will require layering multiple strategies in particular configurations across varying temporal and spatial scales. The guidelines presented here should be integrated with other resource-management considerations, such as phasing of projects across time, land-ownership, permitting, engineering requirements, and monitoring.

DESIRED ECOLOGICAL FUNCTIONS

The process-based strategies are designed to work together to support desired ecological functions in the future Delta: recovering lost support for native species and helping them persist in a changing environment. The ecological functions were chosen to reflect desired support for native wildlife that has been degraded over time (i.e., life-history support for native fish, marsh wildlife, riparian wildlife, waterbirds, and terrestrial species around the Delta's periphery), a productive food web, and overall native biodiversity. We illustrate at two different scales how these functions could be restored: conceptual maps of landscape configurations at the Delta scale, and schematics of how the process-based strategies fit together at a more localized scale.

Several key ideas emerge from the research and synthesis of the Delta Landscapes project that could guide next steps for future planning efforts:

- Different actions are appropriate in different places; therefore, regional visions are a key next step.

- Process-based restoration is a goal for self-sustaining ecosystems, but management will be required.

- Actions should support multiple species and ecological functions.

- Restoring at large spatial scales is critical for success

- Restoration will take time.

- Ongoing learning and adjustment are critical

- Success is attainable

As ecological restoration moves forward, the ideas in this report will need to be integrated with social and economic concerns in stakeholder-based planning processes for different regions of the Delta. As restoration gains momentum, monitoring and adaptive management will be critical for learning as much as we can, as quickly as we can about how to efficiently and effectively regain desired ecological functions within the working landscapes and novel ecosystems of the Delta. The current gaps in our scientific understanding of how the Delta functions, how restoration will affect ecosystems, and how future change will influence the Delta landscape should continue to be addressed through research and well-coordinated adaptive management. However, some uncertainties will only be tackled by moving forward with pilot projects and experimental management actions. Over time, regular evaluation of project goals and accomplishments can keep restoration on track by addressing trade-offs and making adjustments for new information. Despite the many challenges the Delta faces, there is great potential to regain a healthy ecosystem that supports native wildlife while retaining the local culture, agricultural economy, and water-supply services that so much of California relies upon.

INTENDED USE

This report is a guide for resource managers, planners, local governments, and other decision makers who are working to integrate the protection, restoration, and enhancement of Delta ecosystems with agriculture, water management, and other uses. Developed as a technical resource using the best available science, this report is not a policy document. The recommendations can be used by individual agencies through their own particular processes.

Major goals of this report are to:

- Guide restoration planning and design at regional, landscape, and project scales

- Inform stakeholder planning and visioning processes

- Track at the regional scale how local projects are adding up to larger landscapes, and provide advice for optimal, value-added outcomes

- Guide restoration funding priorities

CONTENTS

1

The INTRODUCTION provides the context and setting for this report, as well as broad background on the historical ecology and the major ecological changes that have taken place since the historical period

2

In GUIDING PRINCIPLES, we present general recommendations that apply across the Delta, and should be considered for every project. These are drawn from work by SFEI and other regional scientists and planners, and provide a broad background to the approaches detailed in later chapters.

3

CONCEPTUAL MODELS OF PHYSICAL AND ECOLOGICAL PROCESSES presents a concise overview of key physical and ecological processes in the historical and modern Delta. These map-based schematics show how the physical processes that created a dynamic setting for a diverse ecosystem in the Delta have been drastically altered, leading to less support for desirable ecological processes today.

4

5

6

01
INTRODUCTION

This report offers guidance for creating and maintaining landscapes that can provide desired ecological functions for decades to come. Based on extensive research into how the Sacramento-San Joaquin Delta used to function, how it has changed, and how it is likely to evolve, we make recommendations for how to re-establish the dynamic natural processes that can sustain native Delta wildlife as healthy populations into the future. The approach, building on work others have piloted and championed, is to restore or emulate natural processes where possible, establish an appropriate configuration of habitat types at the landscape scale, and use multi-benefit management strategies to create a more viable Delta ecosystem that can adapt and continue to provide valued functions as the climate changes. This approach is designed to integrate with the human landscape: ecosystem improvements as a part of a robust agricultural economy, water infrastructure and diversions, and urbanized areas. Strategic restoration which builds on the history and ecology of the region can contribute to the strong sense of place and recreational value of the Delta in the future.

Sandhill cranes, Cosumnes River Preseve, 2014, photograph courtesy Bob Wick (BLM)

Isleton water tower, photograph by Kate Roberts (SFEI)

The Delta is a place for people and wildlife

The Delta has long been inhabited by people and wildlife, and people have always valued the Delta for its abundant ecological resources. Indigenous Californians were already here when the tides first began to spread into the Central Valley around 3,000 years ago. As the Delta formed, tribes inhabited many of the tidal islands, likely managing the wetlands with controlled fire, hunting tule elk and waterfowl, and fishing for salmon, Sacramento perch, and other fish. More recently, towns were built on the natural levees of the Sacramento River, and an agricultural economy grew from the tremendous fertility of the Delta's wetland soils. This way of life also centered on hunting, fishing, and boating, benefiting from the robust and productive ecosystem. Despite all the modifications, the Delta way of life, the pace and feel, still comes from the unique geography of the Delta – the broad, slow sloughs, the hidden coves, the tight river bends.

At this moment in time, people of the state of California, and resident and migratory wildlife continue to rely on the Delta for food, water, and recreation, yet the ecosystems that historically flourished and supported these multiple benefits have been critically compromised. Endangered species have declined precipitously,[1] water supply allocation and infrastructure is tenuous,[2] and the risk of catastrophic levee failure is high.[3] Strategic restoration of ecological health can be efficient, can integrate working landscapes of the Delta, and can reinforce the strong sense of place and history in the Delta, in addition to providing ecosystem services and economic benefits. The Delta's future is of statewide significance. An integrated Delta landscape that weaves together history, ecology, and agriculture can increase the region's visibility as an essential resource and a unique treasure that warrants investment for a healthy and viable future as the climate changes and California's population increases.

The Delta is a complicated place

Designing and implementing the appropriate ecological restoration is complex. The Delta serves many purposes: supplying freshwater to a large portion of the state of California, supporting a robust agricultural economy and culture, and providing habitat for native plants and animals. For decades, agencies, stakeholders, scientists, and planners have attempted to create a shared vision for the future Delta,[4] and still the Delta continues to teeter in a precarious situation, hampered by the many seemingly intractable challenges that complicate this landscape.

These challenges include poor water quality, an over-allocated water supply, decaying infrastructure, decline of ecosystems, invasions of alien species, and a complex management structure. We describe some key challenges and uncertainties below:

- The Delta supplies freshwater to 25 million people and three million acres of farmland in southern California, the South Bay Area, and the San Joaquin Valley through pumping facilities in the South Delta.[5] Additionally, water is diverted upstream on the major tributaries to meet Sacramento and San Joaquin valley urban and agricultural needs, and within the Delta for use on its more than 220,000 ha (550,000 acres) of cultivated land.[6] The ability of the Delta to meet these water demands is limited by the need to maintain inflows and outflows to support a healthy ecosystem and the native species

North Delta, photograph by Shira Bezalel (SFEI-ASC)

that depend on it. The challenge of the ongoing drought has further constrained the ability to meet these competing needs.

- More than 1,700 km of levees maintain the Delta's hydraulic integrity. These levees are at risk of failure from ongoing subsidence, flooding, sea-level rise, wind-wave erosion, mammal burrows, and earthquakes. They are mostly privately owned, not professionally engineered, and have limited maintenance.[7] Levee failure results in the flooding of subsided Delta islands, which in turn affects the volume and range of the tides. Such failures can draw in salty water, adversely affecting Delta agriculture and drinking water quality. Additionally, the failure of one levee increases the risk that neighboring levees will also fail, potentially cascading into catastrophic levee failure, with major implications for ecosystems and water quality. While it is highly likely that levees will fail over the coming decades in the absence of concerted action, the timing and size of these failures is unknown. There are several major seismic faults near the Delta that could cause earthquakes able to damage levees.[8]

- Invasive species are likely to continue to affect Delta ecosystems, although the timing, kinds, and ecological implications of new species introductions are difficult to predict. Already, there have been large-scale changes in the Delta from invasive clams, fish, and submerged or floating aquatic vegetation (SAV/FAV). Likely future invaders include the quagga and zebra mussels, which have wreaked havoc on many other ecosystems in eastern North America, damaging native wildlife populations and boat navigation.[9] Climate change is likely to facilitate new invasions.[10]

- Climate change will increase the range of future possibilities. These uncertainties will be considered in the update of the Bay Delta Water Quality Control Plan, and through the requirements of the Delta Reform Act to meet the co-equal goals of improving statewide water supply reliability, and protecting and restoring Delta ecosystems. California WaterFix, the current plan to change the way water is conveyed through the Delta, would divert water from the Sacramento River in the North Delta through two 35-mile-long tunnels during periods of high outflow and reduce reliance on the south Delta pumping facilities, thereby restoring more natural flow patterns in the winter and spring.[11]

Through all of these challenges and uncertainties, change is the constant. The Delta's climate is changing, and increases in air and water temperature, sea level, and the severity of storms and droughts are very likely, among other, less certain effects.[12] These changes will impact flood risk, water supply, human health, agriculture, and ecosystems. Climate-change projections for the Delta watershed include more precipitation falling as rain and earlier snowmelt in the Sierra Nevada. These shifts will greatly affect the annual hydrograph, causing higher peaks earlier in the year and lower flows later in the spring and summer. Based on the state-wide guidance for sea-level rise (SLR) planning, which is now several years old, sea levels are projected to rise 70–185 cm by 2100 in California.[13] Sea-level rise will likely increase the size of the estuary, cause intertidal habitats to migrate or drown, increase salinity levels in the Delta, and add to the pressure on levees.[14] Sea-level rise is projected to accelerate shortly after mid-century, and changes in water level may well be abrupt, rather than continuing its slow and steady increase.[15] Many other aspects of climate change are less certain, and there are likely to be surprises over the coming decades. For example, climate-change models do not agree on how the total amount of precipitation may change.

Because of these uncertainties, it is important to support diverse landscapes that can adapt and be resilient to a range of anticipated and unanticipated perturbations. This motivates research, planning, and action, and monitoring. Recent State policy has set ambitious goals to restore the health of Delta ecosystems. The Delta Plan and Central Valley Basin Plan, as well as other regional documents, have identified the need to go beyond small-scale habitat restoration to create larger landscapes of interconnected habitats that provide desired ecological functions. The new Delta Conservation Framework, in development by the California Department of Fish and Wildlife (CDFW), aims to coalesce decades of Delta science and planning by focusing on, among other things, a long-term continuation of California EcoRestore, initiating a forum for collaborative engagement and broad buy-in with the Delta stakeholder community, providing guidance on conservation project prioritization across space and time, incorporating adaptive management and research, helping to pinpoint key funding priorities for conservation and multi-benefit solutions, and informing the amendment of the ecosystem elements of the Delta Plan.

At present, however, there is a gap between the science that has been completed in the Delta and these ambitious goals set forth in the plans. A paucity of large-scale, long-time-frame, or quantitative guidance hinders efforts to design the complex landscapes and natural systems that are likely to achieve these goals.

Our approach

A Delta Renewed attempts to inform and contribute to ongoing planning efforts by providing a science-based, big-picture perspective on how to re-establish a landscape that functions well for people and wildlife. We offer regional recommendations and on-the-ground strategies to help contextualize, design, implement, and manage future Delta landscapes that can support desired ecological functions, like healthy native fish populations, over the long term. Our approach was to evaluate the landscape patterns and processes that supported wildlife in the historical Delta, measure how they have changed, and assess the potential for establishing smaller, modified landscapes in the future Delta that are productive, sustainable, and supportive of native wildlife. This means creating dynamic systems with the ability to respond to disturbance and stressors in a way that maintains high levels of biodiversity and favorable ecological function. This can be achieved through larger restoration actions, restoring and emulating large-scale natural processes, maintaining complexity and appropriate

Assumptions and Limitations

- This is a science document intended to inform restoration of desired ecological functions in the Delta.

- We assume the Delta will remain as agriculture or developed land. The goal is not to restore the historical Delta.

- Successful restoration of ecological health in the Delta will depend on a well-coordinated and collaborative approach with Delta residents, as well as agricultural and other stakeholders. Such an approach should be the focus of subsequent efforts.

- This document is intended to be used as one resource among many as communities develop regional and local landscape plans. Other key issues are economic constraints, landowner decisions, land-use planning, and societal priorities. It is hoped that the ecological visions, principles, and recommendations herein help inform such planning processes.

- We do not focus particularly on ecosystem services, such as carbon storage or water-quality improvements. However, the Delta does provide many critical ecosystem services that are compatible with our recommendations. For example, estuarine wetlands reduce flooding by attenuating waves and spreading out and slowing down high water, enhance water quality by filtering out and breaking down contaminants, provide nurseries for fish and shellfish, sequester carbon, and provide important recreational opportunities. While it is not the focus of this report, wetlands make valuable contributions to the local economy and quality of life.

connectivity within and among projects, better integrating wildlife support with other land uses, and planning for longer time horizons.

Future change threatens to continue to transform the Delta in undesired ways. Directed action is needed to create a landscape that can provide desired ecological functions and ecosystem services under a variety of future scenarios. Our approach focuses on emulating the large-scale physical and biological processes that can provide such ecological resilience over long time frames. This approach is based on the idea that landscapes with appropriate physical and ecological processes, whether natural or managed, can adapt and evolve in the face of disturbance events and long-term change, thus providing desired functions (such as wildlife support or carbon sequestration) as the climate changes and other events unfold.

The critical physical processes, notably naturalistic flows, beneficial flooding, and the transport of materials and energy they provide, occur at large spatial scales. So do many of the ecological processes, like wildlife migration, primary production of sufficient magnitude to support viable wildlife populations, and the maintenance of sufficient genetic diversity to allow adaptation to future change. To regain these desirable processes, the landscape must function coherently as a whole, rather than as isolated habitat patches. Also, many processes act over long time scales to create and maintain the landforms that support habitat mosaics, and to evolve into new habitats as conditions change. To match the inherent scale of these processes that confer resilience, Delta planning and restoration must also occur at large spatial scales over long time frames.

NOT RECREATING THE PAST
We envision a renewed Delta that incorporates knowledge of the past and present but does not look like either. Given the multiples uses of the Delta, societal and political stressors, invasive species, and future challenges, such as sea-level rise and flooding,

the Delta is and will be a novel ecosystem.[16] The challenge will be to work in concert with underlying topographic and hydrological attributes to recover desired ecological functions, with the goal of maintaining and enhancing native species, as part of the larger working landscape. The future Delta will likely continue to support a range of habitat conditions, including native and non-native species.

RESTORATION TAKES TIME AND ONGOING MANAGEMENT

Restoration actions will result in long stages of interim landscapes with the complexities that operation in the real world brings. For example, land may be acquired slowly over decades. Individual restoration projects should fit within a larger vision, even if all of the pieces of the puzzle can not be assembled at once. The shallow flooding of subsidence reversal projects may, in the short term, have trade-offs for native species, but in the long term may provide much needed connectivity between marsh patches (see Appendix A for discussion of marsh teminology). While the goal is process-based restoration, we acknowledge that in our modified landscapes, self-sustaining processes may not always be possible. Some places will need to be heavily managed, but a focus on reintroducing processes, and not creating static landscapes, will hopefully decrease management intensity.

NEED FOR ADAPTIVE MANAGEMENT

The complexity of the Delta system, combined with future change and uncertainty, requires monitoring and adaptive management of the actions proposed here, changes to policies and permitting, and continued evaluation of project goals and accomplishments over time. Many gaps remain to be filled in our understanding of how the Delta functions, how restoration actions will influence such functions of the Delta, how climate change will impact the system, and much more. A list of Science Gaps and Uncertainties can be found in Appendix C.

Reader's guide

HOW TO USE THIS REPORT

This report is organized into five major chapters. Chapters 1–3 provide background to support the more specific recommendations provided in Chapters 4 and 5. Readers interested in guidance on how to implement process-based strategies, with landscape specifications by habitat type, should focus on Chapter 4. Readers interested in understanding landscape-scale configurations and recommendations to provide support for specific ecological functions at the whole-Delta scale should focus on Chapter 5.

Intended use

This report is a guide for resource managers, planners, local governments, and other decision makers who are working to integrate the protection, restoration, and enhancement of Delta ecosystems with agricultural, water management, and other land uses. Developed as a technical resource using the best available science, this report is not a planning document.

The recommendations can be used to inform planning, restoration, and management of the Delta by individual agencies through their own particular processes. Major goals of this report are to:

- Guide restoration planning and design at regional, landscape, and project scales

- Guide restoration funding priorities

- Inform stakeholder planning and visioning processes

- Track at the regional scale how local projects are adding up to larger landscapes, and provide advice for optimal, value-added outcomes

The report outlines a broad suite of actions that are intended to be evaluated and implemented voluntarily, incrementally, and cautiously in the coming decades. These actions can be adapted to create regional and site-specific solutions that match the particular context and needs of the communities involved. This report relies heavily on the work of many scientists and the many plans proposed for Delta management, such as Envisioning Delta Futures, The Delta Plan, Bay-Delta Conservation Plan (BDCP), and others.

CONTEXT

This final phase of the Delta Landscapes Project builds on previous research and analysis described in two reports: *Delta Historical Ecology Investigation*[17] and *A Delta Transformed*.[18] The former provided a detailed picture of the historical ecology of the Delta, and the latter quantified landscape change by comparing the historical (early 1800s) and modern Delta. See below for more detailed descriptions of both reports.

As the culmination of this project, this report envisions a future Delta, and, therefore, incorporates resilience literature and process-based restoration strategies not discussed in the previous reports. However, the same ecological functions and landscape metrics from *A Delta Transformed* are used as the scientific basis of the proposed strategies.

SCIENTIFIC REVIEW

The challenging task of defining landscape-scale Delta ecological functions, identifying and quantifying landscape metrics, and generating restoration principles and strategies necessitates the collective knowledge and best professional judgment of a team of experts. For this reason, an interdisciplinary group of scientists was assembled to provide guidance and review throughout the Delta Landscapes Project. This group is referred to as the "Landscape Interpretation Team" (LIT) and was drawn from relevant fields of expertise (including geology, geomorphology, hydrodynamics, animal ecology, plant ecology, landscape ecology, and water resource management), many of whom have served on the LIT since the project began in 2012 (see table opposite). LIT members have been consulted individually throughout the project and have met in plenary on eight occasions. The LIT has reviewed and provided input on every phase of this report from initial concepts and approaches, to particular opportunity areas, graphics, and restoration strategies. The recommendations in this report reflect the technical expertise of the LIT but do not necessarily reflect the positions of the agencies for whom they work.

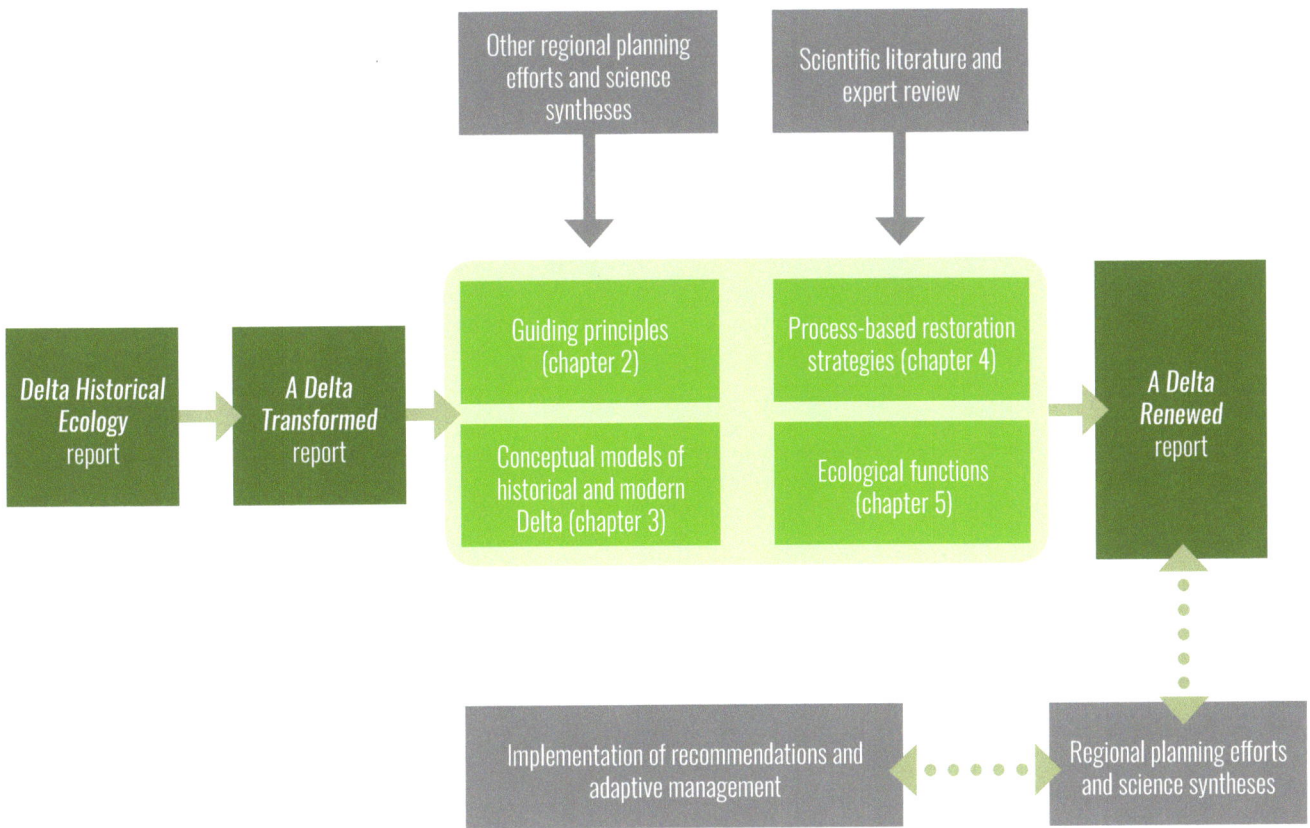

Flow diagram illustrating how elements of the Delta Landscapes project can lead to on-the-ground restoration.

LIT member	Affiliation
Stephanie Carlson	University of California, Berkeley
James Cloern	U.S. Geological Survey
Brian Collins	University of Washington
Christopher Enright	Delta Science Program
Joseph Fleskes	U.S. Geological Survey
Geoffrey Geupel	Point Blue Conservation Science
Todd Keeler-Wolf	California Department of Fish and Wildlife
William Lidicker, Jr.	University of California, Berkeley (Professor Emeritus)
Steve Lindley	National Oceanic and Atmospheric Administration/National Marine Fisheries Service
Peter Moyle	University of California, Davis
Anke Mueller-Solger	U.S. Geological Survey
Hildie Spautz	California Department of Fish and Wildlife
Alison Whipple	University of California, Davis
Dave Zezulak	California Department of Fish and Wildlife

LIT members were essential to guiding and vetting concepts in the Delta Landscapes project.

A short primer on the historical Delta landscape
(summarized from Whipple et al. 2012)

The Sacramento-San Joaquin Delta historically served multiple physical and ecological functions. It was a perennial freshwater source in a Mediterranean climate, collecting, draining, and mixing water from the interior of the state (40% of the state's freshwater flows) to the ocean. It likewise served as an extended fluvial-tidal interface, with tidal influence extending past Sacramento. Saltwater influence was historically limited to the brackish Suisun marshes, and diminished towards Sherman Island, though the boundary was variable depending on the year. Unlike most deltas which spread out, the Sacramento-San Joaquin Delta has an inverted shape, narrowing at its outlet before opening to the San Francisco Bay.[19] It functioned as a sediment sink, slowing and settling coarser materials eroded from the granitic Sierras, while passing sands

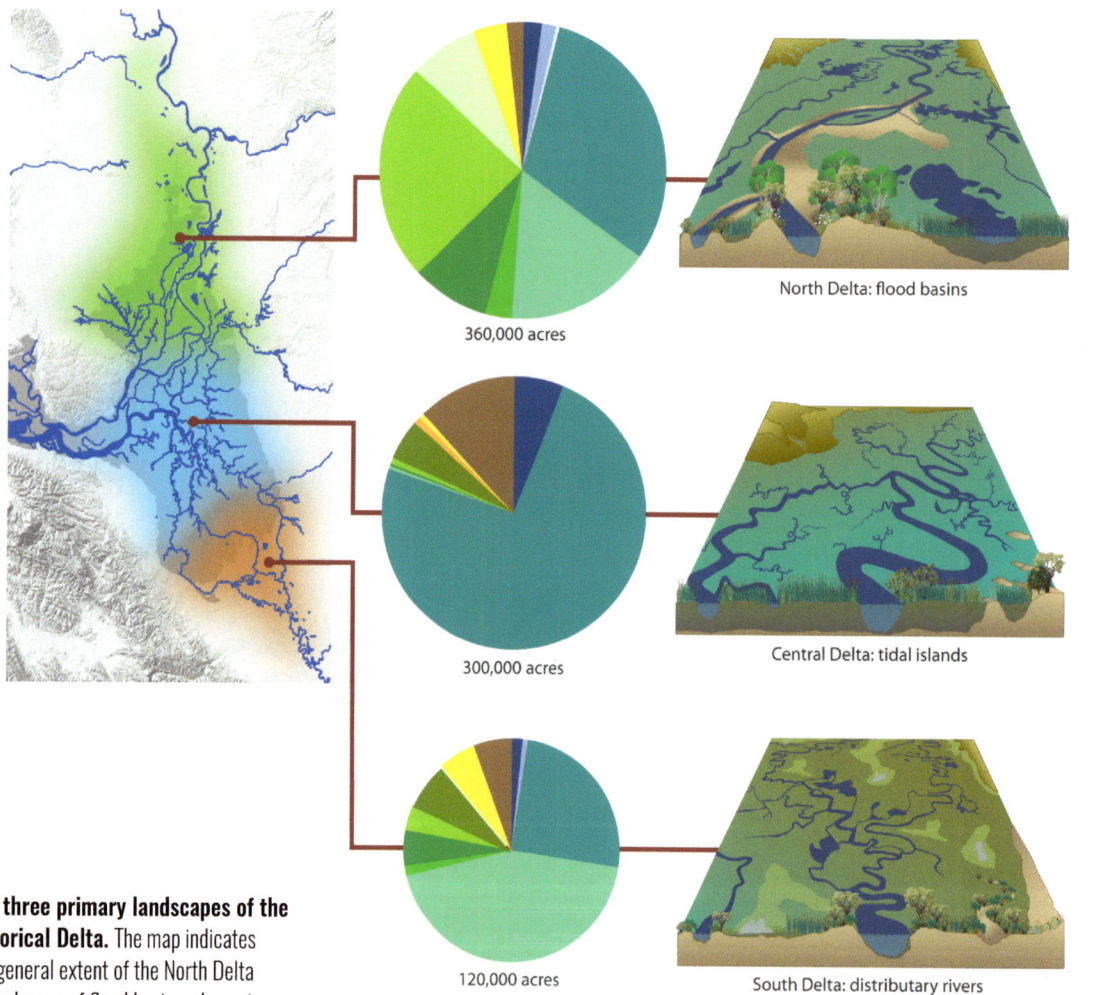

360,000 acres

North Delta: flood basins

300,000 acres

Central Delta: tidal islands

120,000 acres

South Delta: distributary rivers

The three primary landscapes of the historical Delta. The map indicates the general extent of the North Delta (a landscape of flood basins; shown in green), Central Delta (a landscape of tidal islands; shown in blue), and South Delta (a landscape of distributary rivers; shown in brown). These landscapes were characterized by different assemblages and relative proportions of habitat types (as seen in the pie charts). Conceptual diagrams illustrating each of these landscapes are shown to right.

Water
Pond/lake
Seasonal pond/lake
Tidal freshwater emergent wetland
Non-tidal freshwater emergent wetland
Willow
Valley foothill riparian

Wet meadow and seasonal wetland
Vernal pool complex
Alkali seasonal wetland complex
Stabilized interior dune vegetation
Grassland
Oak woodland or savanna

and silts downstream to replenish the salt marshes and beaches. It was also the lungs of the region, sequestering carbon and releasing oxygen. The Delta was a highly productive system that provided abundant and diverse food resources to support robust food webs, as well as indigenous tribes. Many native wildlife species were able to exploit the complex and resource-rich landscape of the Delta, some thriving in astonishing numbers.

The historical reconstruction of the Delta reveals the large-scale patterns and heterogeneity that existed before major anthropogenic influences.[20] The central, northern, and southern parts of the Delta were diverse in their geomorphic and hydrologic settings, and in the ecological functions they provided. The Central Delta consisted predominately of islands of tidal freshwater emergent wetland (marsh), which supported a matrix of tule, willows, and other species. These wetlands—topographically almost flat—were wetted by twice daily tides, and inundated monthly (if not more frequently) by spring tides. During high river stages in the wet season, entire islands were often submerged under several feet of water. The large tidal sloughs had low banks and, like capillaries, bisected into numerous, progressively smaller branching tidal channels which wove through the wetlands, bringing the tides onto and off of the wetland plain, promoting an exchange of nutrients and organic materials. Channel density in the Central Delta was greater than in the less tidally dominated northern and southern parts of the Delta (but lower than in the brackish and saline marshes of the estuary downstream). The edges or transition zones around the Central Delta were composed of alkali seasonal wetlands, grassland, oak savannas, and oak woodlands. On the western edge of the Central Delta, sand mounds (remnant Pleistocene dunes) rose above the marsh, providing gently sloping dry land in an otherwise wet landscape that served as a high-tide refuge for terrestrial species.

The ecological functions provided by the North Delta were driven primarily by the great Sacramento River, which created large natural levees and flood basins. These flood basins, running parallel to the river, accommodated large-magnitude floods, which occurred regularly, with inundation often persisting for several months. They consisted of broad zones of non-tidal marsh that had very few channels and transitioned to tidal wetland towards the Central Delta. Dense stands of tules over 3 m tall grew in these basins. Large lakes occupied the lowest points in these flood basins.

The North Delta's natural levees, created pre-Holocene by the large sediment supply of the Sacramento River, were broad, sloping features that graded into the marsh. These supra-tidal levees supported dense, diverse, multi-layered riparian forests often up to 1.5 km in width. They ran parallel to the Sacramento River and other large tidal sloughs that conveyed enough sediment to build them over time during high flow events. The levees provided migration corridors for birds and mammals, organic debris, and shade to the river systems for aquatic species. Some areas within tidal elevations were seasonally isolated from the tides due to the presence of these levees and complex fluvial and tidal interactions. At the edge of the North Delta, willow thickets occupied the "sinks" where smaller rivers and creeks spread into numerous distributary channels and dissipated into the adjacent wetlands of the floodbasins. Other parts of the North Delta were lined with vernal pool complexes and other seasonal wetlands.

The South Delta, like the North, was shaped by a large river system. Here, the three main distributary branches of the San Joaquin River created a complex network of smaller distributary channels, oxbow lakes, tidal sloughs, and natural levees of varying heights which graded across the long fluvial-tidal transition zone. In contrast with the single main channel of the Sacramento and the parallel flood basins, the San Joaquin River had less power and sediment supply to build high natural levees, and thus had many channels branching from the mainstem and coursing through the marsh islands; these channels vacillated between being fluvially or tidally dominated, depending on the time of the year. Small lakes and ponds were scattered in the South Delta, and the marsh was intersected with willow thickets, seasonal wetlands, and grasslands, making it a very diverse place for wildlife. The edge of the South Delta was dominated by alkali seasonal wetland complex, grassland, and oak woodland. The eastern edge of the Delta was shaped by the alluvial fans of the Mokelumne and Calaveras rivers that spread into the marsh.

The Delta was not a static place. Though the positions of large tidal channels, natural levees, and lakes were relatively stable, the Delta would have looked very different depending on the year and season. Areas of marsh that were flooded with several feet of water by late winter could be dry at the surface by late fall. The Delta was a place of significant spatial and temporal complexity at multiple scales.

A short primer on A Delta Transformed
(summarized from SFEI-ASC 2014)

A Delta Transformed quantifies change in the Delta from the early 1800s to the early 2000s using landscape metrics, including channel length, marsh patch size, riparian width, and extent of seasonally inundated habitats. In total, thirteen metrics are detailed in the report. These metrics were designed to directly bear upon important ecological functions that supported native species.

The defining characteristic of the historical Delta was its extensive wetlands, formed over time as rivers and tides met and spread out across the flat landscape. Modern land management has increasingly disconnected floodwaters from the wetlands by widening and deepening channels, diking and draining wetlands for agriculture, and building levees for flood protection. The consequences of this disconnection include a near complete loss of Delta wetlands, along with the processes that sustain them, and a dramatic altering of the remaining aquatic habitats. The quality of remaining habitats within the Delta has been degraded by a loss of complexity and the addition of anthropogenic stressors.

The habitats that dominated when the Delta was a functionally intact ecosystem have been reduced to small fractions of their former extent. This decrease has reduced the population viability of native wildlife by reducing the size, variability, and connectivity of many populations. The reduced extent of high-endemism habitats, such as vernal pools and alkali wetlands, may have significant consequences for biodiversity in the region. As a result of the diking of marshes, dendritic channel networks have been lost, with ecological consequences for native fish. The loss of high-productivity marsh and floodplain habitats has reduced the food resources available for fish and water birds. Historically there was considerable geomorphic and hydrological heterogeneity within Delta habitats, creating diverse options for wildlife. The modern Delta has lost connectivity within and among habitat types, with the exception of large channels, which are now over-connected, reducing in-channel heterogeneity and altering flows.

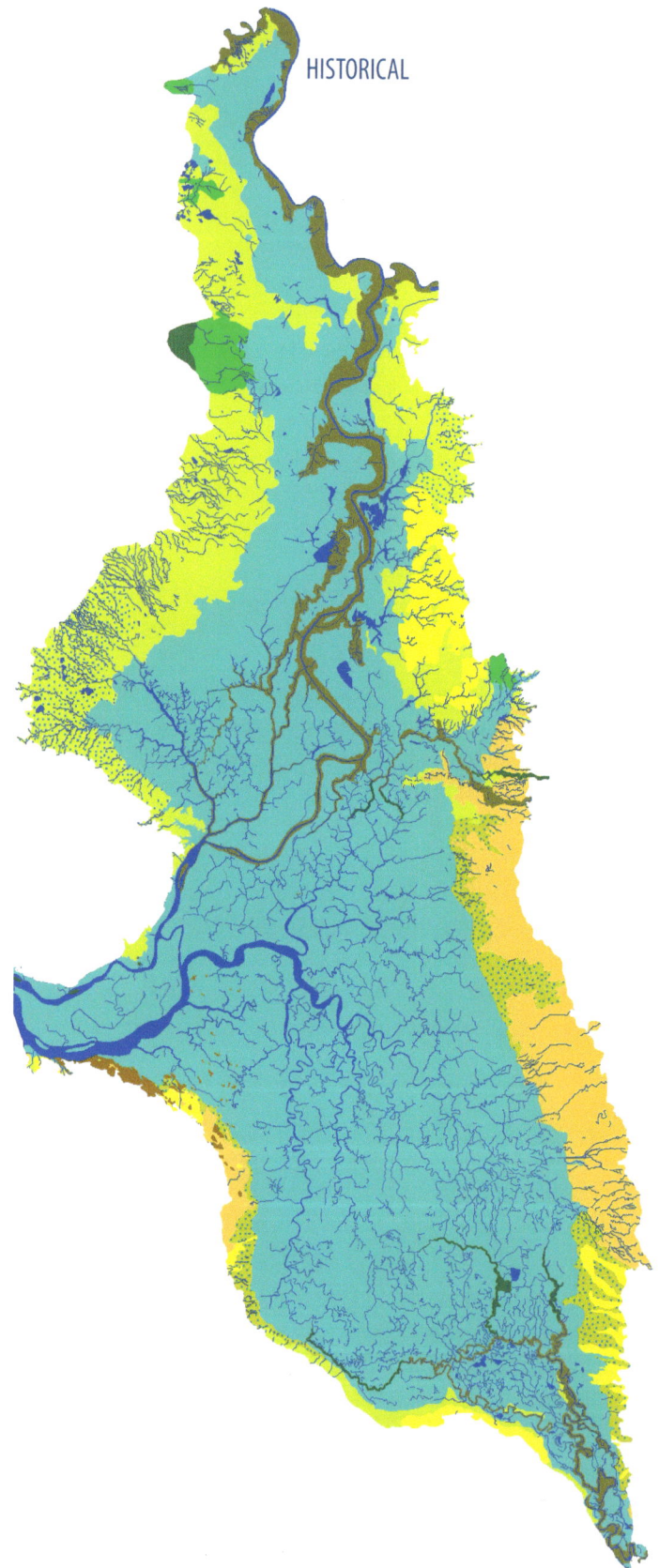

HISTORICAL

MODERN

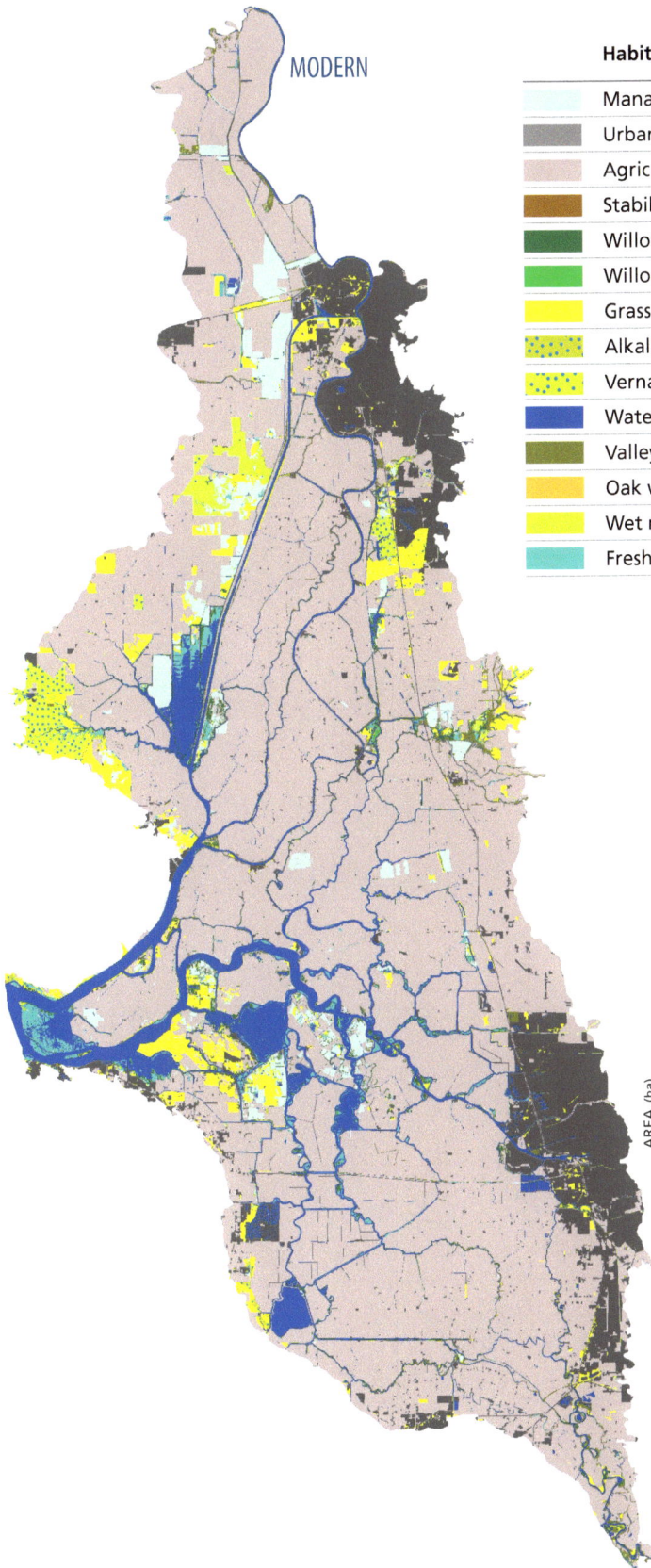

Habitat Type	Area (ha)		% Change
	Historical	*Modern*	
Managed wetlands	0	9,454	∞
Urban/Barren	0	35,517	∞
Agriculture/Non-native/Ruderal	0	216,085	∞
Stabilized interior dune veg.	1,032	4	-99
Willow riparian scrub/shrub	1,637	2,878	+76
Willow thicket	3,567	132	-96
Grassland	9,108	11,800	+30
Alkali seasonal wetland complex	9,193	238	-97
Vernal pool complex	11,262	3,007	-73
Water	13,772	26,530	+93
Valley foothill riparian	15,608	4,010	-74
Oak woodland/savanna	20,460	0	-100
Wet meadow/Seasonal wetland	37,561	2,445	-93
Freshwater emergent wetland	193,224	4,253	-98

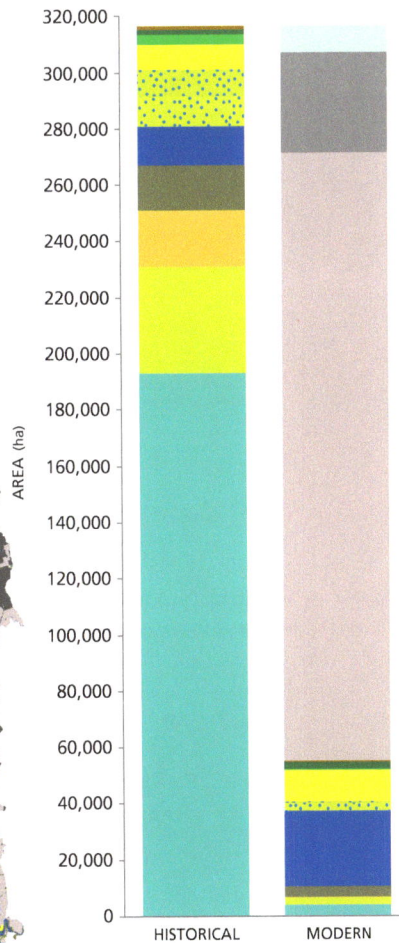

Habitat change. The extent of wetland habitats has decreased in the modern Delta while the extent of open water and grasslands has increased. Agriculture and managed wetlands make up a large portion of the modern Delta and provide some important wildlife support but are not equivalent to historical habitats. Oak woodlands and interior dune scrub have mostly been eliminated.

02 GUIDING PRINCIPLES

16

These guiding principles are general considerations that apply across conservation planning, restoration, and management activities in the Delta. The goal of these principles is to maximize desired ecological functions both in the short term and over long time frames. The principles draw from several recent efforts to develop science-based approaches to achieving long-term ecological health and resilience for the Bay-Delta system. They draw from the Landscape Resilience Framework[1] which, based in the literature on ecological resilience, offers principles that should be applied in tandem to realize large-scale ecological benefits. The principles also include ideas and concepts from the Baylands and Climate Change report,[2] the Delta Plan,[3] and other regional efforts. These principles apply Delta-wide and they underpin the more specific recommendations in Chapters 4 and 5.

River Road near Isleton, photograph by Kate Roberts (SFEI-ASC)

1. Appreciate that people are part of the Delta,

including the individuals, communities, and institutions that steward landscapes. People can build ecological resilience through stewardship of the land, community engagement that broadens the base of support for environmental improvement, and adaptive management that enables responsiveness and flexibility. In the Delta, considerations related to people, institutions, and infrastructure that are critical for successful restoration and conservation include

Ecological, social, and economic resilience. Integrate the recommendations in this report with economic and social priorities. The economic and agricultural resilience of the Delta are critically important. However, these aspects of resilience are not the subject of this report. Rather, this report provides guidance for improving the resilience of desired ecological functions in the Delta. Ideally, planning should aim to maximize ecological, social, and economic resilience.[4]

Developed and agricultural lands. Take advantage of opportunities to support valuable ecological functions in urban and agricultural lands that will not be restored to more natural ecosystems.[5]

Policy and planning. Coordinate policy, planning, research, and monitoring for ecosystem health at the full Delta scale. This includes coordination among policies, projects, institutions, human communities, and individuals.

Adaptive management. Recognize the inherent uncertainty in our understanding of ecosystem responses to change, and build in mechanisms for learning and adaptation that ensure long-term success even in the event of initial failure and other surprises. Adaptive management in the Delta should incorporate opportunities for research and monitoring into conservation actions.

Ongoing management. While some areas will require intensive management, restoration approaches for ecological resilience will increase the likelihood of success and reduce the level of intervention over the long term. Some issues requiring ongoing management are invasive species (e.g., clams, submerged aquatic vegetation, and predatory fish), water temperature, contaminants, and other water-quality concerns (e.g., mercury, nutrient loading, and pesticides).

Infrastructure. Take advantage of infrastructure upgrades to increase ecological resilience, including infrastructure for water supply, stormwater, wastewater, flood control, parks, and utilities. Reduce or eliminate the establishment of hard infrastructure that is not likely to be sustainable in the long term due to flood risk and sea-level rise. Further development in such areas will foreclose on ecological opportunities and will likely be difficult to maintain into the future.

Multiple benefits. When managing for other priorities, such as ecosystem services, integrate ecological benefits to optimize outcomes.[6] For example, current efforts to increase carbon sequestration, improve flood protection and water quality, and expand recreation all provide opportunities to increase ecological functions at the same time. Methods for optimizing outcomes across multiple benefits are not fully developed in all cases, and additional research may be needed.

Stakeholder meetings. Convene stakeholder meetings to integrate Delta-scale recommendations, planning, and policies into local and subregional plans.

Landscape scale. Plan restoration and conservation at the landscape scale to maximize ecological benefits, which will reduce the overall footprint needed.[7] Design restoration projects to minimize potential conflict with agricultural and urban neighbors.

2. Consider landscape context to apply the right strategies in the right places.

Identify key aspects of the geographic setting that are likely to affect restoration and management, and consider how to work with these elements when developing strategies for a site. Understanding the setting or context may require a period of investigation before a project can occur. This research phase may uncover information that enables better matching between conservation actions and the location. Matching appropriate activities to the setting can enhance ecological resilience by focusing actions in the locations where they are most likely to be successful over time, while also reducing cost. Examples of some critical components of setting in the Delta include:

Landscape position. Consider landscape position in the Delta before designing a project. Tidal influence, salinity, soil type, elevation, and local effects of climate-change projections (including sea-level rise, temperature increase, and greater severity of storms and droughts) may all differ by landscape position. Thus, landscape position determines what types of activities will be most appropriate for a site now and into the future.[8] For example, restoration of edge habitat types will not be sustainable in the subsided islands of the Central Delta. Similarly, tidal marsh restoration should occur at sites that are appropriate given sea-level rise projections and plans for sustaining sufficient sediment supply.

Landscape trajectories. Think through likely landscape trajectories and their implications during the planning phase of any project. These trajectories can affect what types of restoration and conservation activities are possible and which are likely to succeed in the future. For example, sea-level rise will change where tidal marshes can be restored in the future.[9] Climate change will alter the timing of flows into the Delta from the Sierra Nevada, and especially from the San Joaquin River. Levee failures and any increase in water diversions will affect which habitat types can be supported where. Consider also the liklihood of self-repair over time contributing to the restoration process.

Biological legacies. Plan projects to take advantage of biological legacies, such as extant and historical wild plant and animal populations (including intact remnants that may contain valuable alleles or social knowledge), seed banks, and snags and downed logs. These legacies affect which plants and animals are likely to benefit from conservation activities, and they can jump-start the restoration process.[10] For example, persistent seed banks can affect which plants will colonize a newly restored area.

3. Restore critical physical and biological processes. These processes are key to maintaining resilient ecosystems that can sustain themselves with minimal human intervention.[11] Examples of critical processes in the Delta include:

Freshwater flows. Restore flows with the necessary frequency, magnitude, duration, timing, and rate of change to create and maintain habitats for native species. Sufficient flows are critical for numerous reasons, including supporting hydrological gradients, providing cues for fish movement, and supporting freshwater marsh by reducing salinity intrusion.

Beneficial flooding. Restore beneficial flooding (tidal and fluvial) of the necessary frequency, magnitude, duration, and timing to support food webs, reproduction, and rearing. For example, floodplain inundation of sufficient magnitude and duration can help support salmonid rearing and splittail spawning.

Tidal Energy. Research is needed to understand how tidal energy will shift with increased tidal marsh restoration in the Delta, sea-level rise, and the possibility of levee failures. Tidal energy drives the maintenance of tidal channels and marshes, and is currently limited in the Delta (thus large restoration projects may effectively dampen tidal range), yet it may also increase with sea-level rise.

Sediment. Plan for how sediment dynamics (transport, deposition, and resuspension) will impact a project, and vice versa. Sediment is a key component of various physical processes in the Delta. For example, sediment deposition enables marsh accretion, and suspended sediment creates turbid conditions that support native fish. Sediment will be an increasingly valuable resource as sea-level rises, and long-term planning for sediment supply is important for Delta restoration.

Primary productivity. Restore multiple primary producer groups and large areas of productive wetland and shallow-water habitat types to provide nutrition for consumers. In the Delta, diversity in the sources of primary productivity is important to support a complex food web. Some examples of producer groups in the Delta include phytoplankton, marsh plants, benthic and epibenthic algae, submerged aquatic vegetation, floating aquatic vegetation, and imported organic matter from woody riparian forest and scrub.[12]

Adaptive evolution. Create landscapes in which natural selection drives the evolution of native wildlife. This will require adequate population sizes and landscape complexity to support sufficient genetic variability and connectivity between small populations. The ability to adapt to changing conditions will be ever more important as climate disruption drives rapid environmental change. At the same time, re-establishing more natural or native-type ecosystems in the Delta will help native wildlife persist by creating an environment they are adapted to.

Sacramento River near Paintersville, photograph by Amy Richey (SFEI-ASC)

4. Restore appropriate landscape connectivity.

Create appropriate connections within and among landscapes, and between aquatic and terrestrial habitat types. Connectivity enables the movement of wildlife and plants, allowing individuals and populations to move in response to stress or to access resources.[13] In some cases, reducing connectivity may be appropriate to increase heterogeneity, allow populations to adapt in isolation, or create redundancy that can enhance ecological resilience.[14] Priorities for restoring appropriate landscape connectivity in the Delta include:

Land-water connections. Restore broad estuarine-terrestrial transition zones to reconnect land and water around the Delta perimeter. Create complex channel networks and remove levees where possible to increase the length of wetland-channel interfaces.

Appropriate connectivity within a habitat type. Restore appropriate levels of connectivity between patches of the same habitat type. Most terrestrial and wetland habitat types are not sufficiently connected, and restoration should enhance or create pathways for the movement of genes and organisms between habitat patches.[15] For example, marshes should be placed close enough to support the gene flow of marsh wildlife (e.g., black rails). Appropriate levels of connectivity may mean reducing connectivity where it decreases desirable habitat heterogeneity (e.g., in artificially over-connected channels).

Connectivity among different habitat types. Restore and manage patches of different habitat types in configurations that support the needs of native biota. For example, seek configurations that preserve and enhance pathways for anadromous fish, including linking marshes and floodplains to channels, or designing for upland-wetland adjacency to support the life-history needs of pond turtles that lay eggs in uplands. Prioritize configurations that increase adjacency between marsh and open water to support the exchange of productivity and nutrients.[16] Maintain the physical processes that facilitate appropriate connectivity between habitat types.

Barriers to movement. Remove or reduce barriers where they negatively impact connectivity, including gaps in woody riparian corridors, and diversions and dams that restrict the upstream movement of anadromous fish using the Delta.

Cosumnes River Preseve, 2014, photograph courtesy Bob Wick (BLM)

5. Restore landscapes with a focus on complexity and diversity. Restore and protect landscapes that include a variety of habitat types and habitat type adjacencies, and that have heterogeneous and complex habitats.[17] Prioritize activities that protect and enhance biodiversity both within and among habitat types. Emphasizing complexity and diversity can promote ecological resilience and enhance native biodiversity by providing a range of options for species, and by expanding the types and numbers of species that a landscape can support. Examples of the types of complexity and diversity that are important in the Delta include:

Variety of habitat types. Restore a variety of habitat types, with habitat-type adjacencies that support desired ecological functions at the landscape scale. Important habitat types in the Delta include tidal and non-tidal marsh, dendritic channel networks, submerged native aquatic vegetation, estuarine-terrestrial transition zones and adjacent uplands, riparian forests, ponds and lakes, and floodplains. In addition, novel habitat types,[18] such as wildlife-friendly agriculture and managed wetlands, can also support wildlife, food webs, and biodiversity.

Complexity within habitat types. Design complexity into restored and novel habitat types as appropriate to the setting, including topographic complexity, physical heterogeneity (e.g., microhabitats and microclimates), and vegetative diversity (e.g., species and structure). In the Delta, some examples of within-habitat complexity include complete channel networks (with first through third or fourth order channels) and ponds and potholes within marshes.

Gradients. Restore physical gradients across the Delta, such as the temperature gradient from north to south, the climate gradient from the cooler inner Delta to the warmer outer Delta, and the salinity gradient from west to east.

Biodiversity. Protect and enhance genetic, species, and ecosystem diversity in the Delta. Manage for species of conservation concern as well as for biodiversity in general. Protect populations of common species as well as rare species; common species often perform important ecosystem functions, including supporting the base of a robust food web.[19] Preserve diversity within species as well; examples include supporting different runs of Chinook and genetically distinct populations of giant garter snake.

23

6. Create redundancy of key landscape elements, populations, and habitat types. Redundancy can increase ecological resilience by providing backups, so that loss of a population or landscape element does not lead to extinction of a species or elimination of a landscape element across the entire Delta.[20] In the Delta, some examples of critical areas to build and enhance redundancy include:

Discrete habitat patches and structures within habitat types. Create redundant habitat patches, and especially increase limiting habitat features. For example, restore multiple large marsh patches. Include, where feasible, wildlife-friendly structures such as tree cavities in woody riparian areas for nesting birds and roosting sites for bats.

Populations. Protect multiple populations of native plants and animals in the Delta (e.g., multiple runs of anadromous fish, multiple populations of vernal pool fairy shrimp).

Multiple movement corridors. Restore and create movement corridors, so that wildlife have options for dispersal. Some critical movement corridors where redundancy should be improved in the Delta are migratory pathways for salmon and woody riparian corridors for birds and other wildlife.

Functional redundancy. Design landscapes to support multiple species performing similar functions,[21] such as redundancy in primary producers, insectivorous animals, granivorous wildlife, and medium to large predators. One important goal is to increase redundancy in sources of primary production from both marsh and aquatic habitat types. Promote redundancy in functional groups that operate at different spatial or temporal scales.[22] For example, both tule elk and California voles are valuable grazers in the Delta.

7. Restore at large scales, with a long time horizon in mind.

Restoration design and conservation planning oriented towards large spatial scales and long time frames will increase the likelihood of creating natural systems capable of sustaining desired functions through an uncertain future.[22] Not all systems and restoration projects in the Delta can be self-sustaining, and intensive management will likely be required in many areas. However, restoring with a long planning horizon will increase the chances of succesful actions that benefits native species over the long term. Key considerations include:

Large habitat patches. Create patches of sufficient size to support key physical processes. For example, tidal marshes should be large enough to support the formation of multi-order dendritic channel networks and self-sustaining wildlife populations.

Long planning horizons. Choose planning time frames sufficient to encompass expected dynamics in physical and biological processes (such as large floods and plant community succession) and to prepare for climate change (such as sea-level rise, shifts in habitat types, temperature increase, and other projected changes).

Linked activities beyond the Delta. Coordinate conservation, restoration, and monitoring to support wildlife that move in and out of the Delta, including waterbirds that migrate along the Pacific Flyway (and use San Francisco Bay, the Central Valley, and the Delta) and anadromous fish that migrate from the upper watershed to the ocean. Planning should be coordinated so that activities in the Delta can benefit activities to restore and manage habitat in other parts of the migratory pathways for these highly mobile species.

Linked activities across scales. Coordinate activities across different spatial and temporal scales. For example, link local and regional-scale management. Restore soon, and yet plan for habitat types to evolve over time. For example, create tidal marshes that can keep up with sea-level rise by allowing them to become established and gain elevation capital now before sea-level rise accelerates. Plan for these marshes to migrate to higher elevations on the Delta periphery as sea-level rises.

Cosumnes River Preserve, photograph by Shira Bezalel (SFEI-ASC)

CONCEPTUAL MODELS OF PHYSICAL AND ECOLOGICAL PROCESSES

Understanding the physical and ecological processes that shaped and maintained the Delta historically, and continue to do so today, is essential to understanding how the system has changed and to identifying the interventions that can restore ecological health and maintain that health in a resilient way. This chapter presents conceptual models of physical and ecological processes in the historical and modern Delta. The conceptual models of the historical Delta show how physical processes created a dynamic template to support complex ecological processes as part of a diverse ecosystem. The conceptual models for the modern Delta show how these physical processes have been altered from their historical dynamics, and how these changes have led to less support for critical ecological processes today. The physical processes described in these conceptual models underlie the process-based strategies described in Chapter 4; the ecological processes described are critical for maintaining the ecological functions discussed in Chapter 5. These simple conceptual models frame the conservation and restoration guidance provided in this report. More detailed models of Delta processes can be found in the Delta Regional Ecosystem Restoration Implementation Plan (DRERIP) conceptual models and Interagency Ecological Program (IEP) tidal marsh workgroup revisions to those models.[1]

North Delta, photograph by Kate Roberts (SFEI-ASC)

Physical processes of the
Historical Delta

At the beginning of the Holocene, a period of climatic warming caused glaciers to melt and sea level to rise rapidly (20 mm/yr), forming many of the features we recognize as the modern Bay-Delta estuary.[2] As sea level rose, tides extended eastward across floodplains and valleys upstream of the Golden Gate. San Francisco Bay began to form 10,000 years ago, and tidal wetlands spread into the site of the Delta by 6,000 years ago.[3] Around that time, the rate of sea-level rise slowed to about 10 mm/yr, allowing tidal marsh accretion to keep pace with the rate of sea-level rise.[3] Over time, the tidal marshes formed thick layers of peat (20 m) in the Central Delta.[4] In general, edges of the Delta had thinner layers of peat (as tidal influence more recently reached these areas) interbedded with fluvial sediments deposited by the major river systems which terminate in the Delta.[5]

Tidal and fluvial processes continued to build and sustain the historical Delta as it existed in the 1850s. The tides maintained an extensive dendritic channel network throughout vast freshwater marshes. The daily tides, along with elevated, storm-induced tides and high river flows, brought an influx of water, nutrients, and organic matter that supported wetland vegetation, peat soil development, and equilibrium in the elevations of the water surface and the marsh plain over time. High river flows also brought large volumes of sediment to the Delta, with coarser sediment depositing at river mouths and adjacent to river channels, forming fans and natural levees that supported distinct vegetation types. Some of the finer sediment deposited on the Delta marsh plain, but most was transported through the Delta and out to San Francisco Bay where it nourished tidal marshes and mudflats. Indigenous people also contributed to the maintenance of Delta habitats by managing marsh and floodplain vegetation with fire.[6]

Beneficial flooding: The alluvial fans of these tributaries defined the Delta's northwestern edge. Though only directly connected to the Delta during times of flood, the creeks delivered sediment and nutrients to the northern part of the Delta, sustaining marshes in that region. Freshwater wetlands and willow thickets (or "sinks") formed at the mouths of smaller rivers and were supported by high groundwater. The topographic gradient along the face of the fan down to the marsh plain resulted in zones with high groundwater and a distinct transition from woody riparian to wetland vegetation.

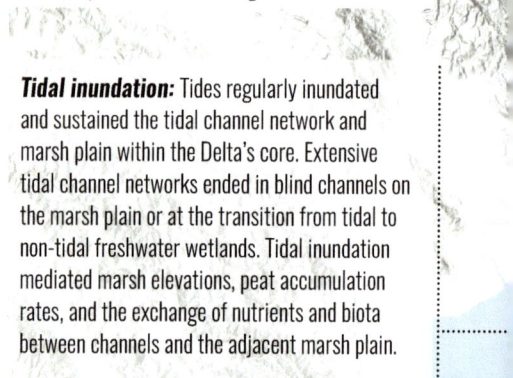

Tidal inundation: Tides regularly inundated and sustained the tidal channel network and marsh plain within the Delta's core. Extensive tidal channel networks ended in blind channels on the marsh plain or at the transition from tidal to non-tidal freshwater wetlands. Tidal inundation mediated marsh elevations, peat accumulation rates, and the exchange of nutrients and biota between channels and the adjacent marsh plain.

Micro-topographic complexity: Sand blown eastward from ice-age floodplains built dune fields in the western Delta. Rising sea levels drowned the sand sources and allowed tidal wetlands to surround the highest dunes. These stabilized dunes became sand mounds that added ecological complexity and influenced the distribution of tidal channels.

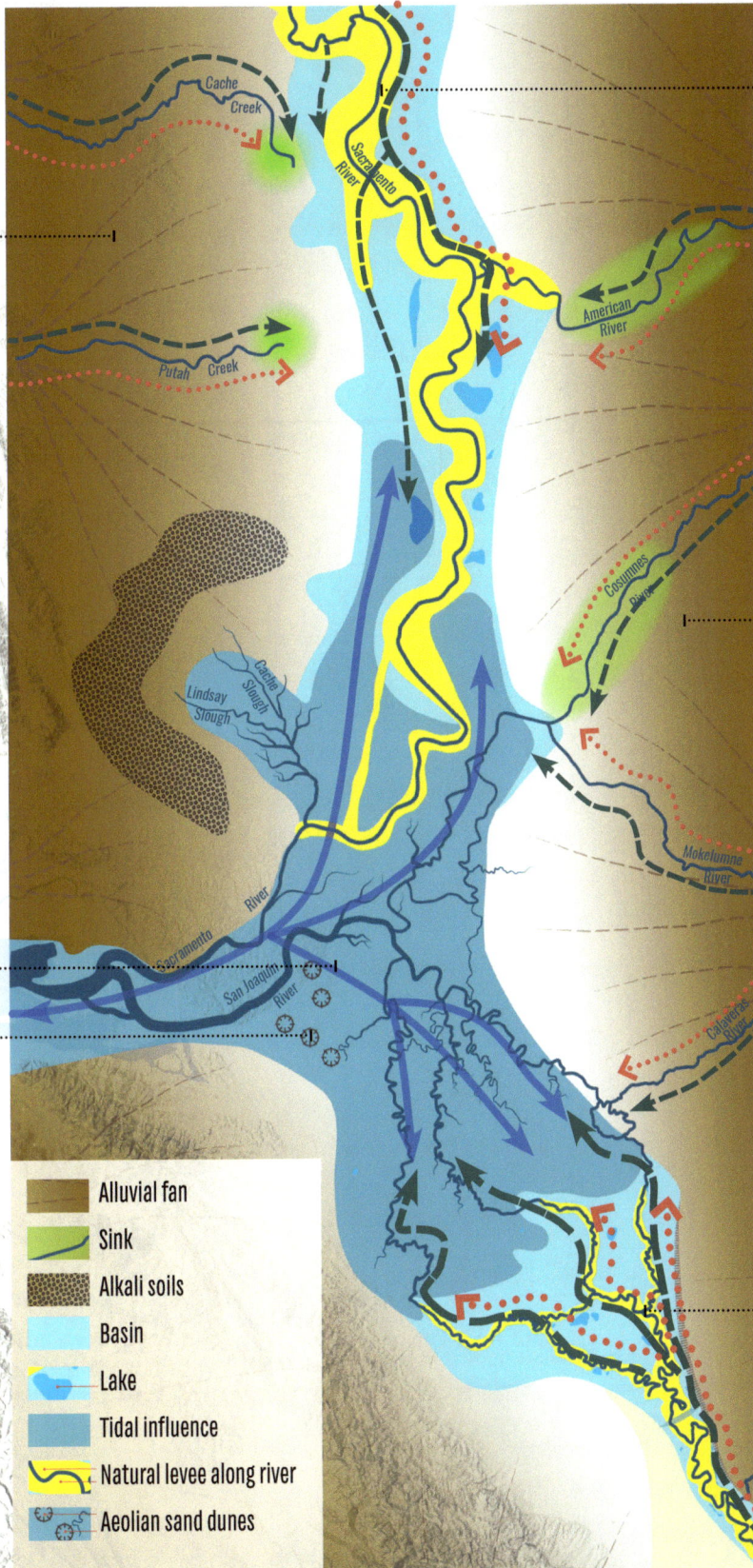

Freshwater and sediment delivery:
The Sacramento River regularly spread out on its expansive floodplain during the winter months, inundating and helping sustain the network of adjacent large basins and freshwater lakes as well as the marsh plain downstream. The river's high sediment load created broad natural levees that supported dense, wide riparian forests, while extreme flows spilling into the Yolo Basin created crevasse splays and long-duration flooding.[7]

Tidal-freshwater gradient:
Winter flood flows from channels draining snow-melt within glaciated Sierra Nevada watersheds inundated and helped sustain the marsh plain throughout the Delta. The alluvial fans that these channels built over time defined the Delta's eastern edge. The creeks formed sinks at their mouths, and the topographic gradient from their fans down to the marsh plain created zones with high groundwater and a distinct transition from woody riparian to wetland vegetation.

Fluvial-tidal gradient: The delta of the San Joaquin River was characterized by distributary channels and a network of ponds and lakes supported by snowmelt flood flows. These flows inundated and delivered water and nutrients to the marsh plain downstream. The relatively low sediment load and lower peak flows (compared to the Sacramento River) resulted in low-elevation natural levees along the distributary channels. As with the Sacramento River, these levees graded down into the marsh plain, tracking the gradient from non-tidal to tidal environments, creating a change in vegetative structure and function.

Legend:

- Alluvial fan
- Sink
- Alkali soils
- Basin
- Lake
- Tidal influence
- Natural levee along river
- Aeolian sand dunes

PROCESS ARROWS

- Tidal flow
- Stream flow
- Sediment delivery

29

Physical processes of the
Modern Delta

Many of the processes that once shaped the Delta have irrevocably altered since the historical period. Marsh reclamation in the 19th century and subsequent farming on the productive peat soils led to peat oxidation and major subsidence in the Central Delta.[8] The large Sierran rim dams, along with other water diversions upstream, decreased the delivery of coarse sediment and changed the amount and timing of flows to the Delta. These changes caused incision of the rivers that enter the Delta, cut off the peak flows that regularly inundated the Delta's floodplains and marshes, and starved the estuary of the sediment needed for marshes and mudflats to persist over the long run. The loss of sediment supply was countered by a large mass of hydraulic mining debris that moved through the system for more than a century. Channels in the Delta have been widened and deepened, leading to more homogenized water conditions (e.g., salinity, temperature, nutrients, velocity) and altered tidal and flood routing through the Delta.[9] The levee network, new channels constructed to increase connectivity, and new flood bypasses move flood flows from the surrounding watersheds rapidly through the Delta and decrease flood risk for urbanized areas.[10] Though groundwater levels in the Central Delta are still generally high, local pumping has caused significant declines at certain locations along the Delta's periphery, especially around cities.[11] These declines likely limit or prevent the establishment of groundwater-dependent edge habitats.

Further transformation of the Delta can be expected as climate changes unfold over this century. Pressure on the levee system will continue to increase with sea-level rise and larger storm events. Tidal marshes may be at risk of drowning as sea level rises, if they are not connected to inorganic sediment sources. Marshes will migrate inland as sea level rises, if given the space. Understanding the trajectory of processes acting on the Delta may help restoration professionals design projects that take advantage of these changes.

Despite the radical changes that have taken place, there is hope for renewing the Delta. Of the several large-scale drivers that helped create and maintain the Delta, many are still either fully or partially intact and can be used to inform restoration concepts that could be successful in improving ecosystem health.

Watershed inputs: These historically disconnected tributaries have been modified to have a permanent connection to the Yolo Bypass and are confined by flood-control levees. Dam building in the tributary watersheds has likely decreased the long-term delivery of flows and sediment to the Delta.

Tidal inundation: Tidal flow is now constrained to large, leveed channels, and the complex, dendritic marsh channels are largely gone. As a result, tidal inundation extent has been reduced drastically, which has impacted land elevations, primary production, and the movement of nutrients and biota throughout the Delta.

Subsidence and peat oxidation: The core of the Central Delta has subsided 1-8 m below sea level across over 60 islands.[12] The cause has been the oxidation of peat deposits, formed over 6,000-7,000 years that once formed a peat layer between 2-15 m thick. More than half of the peat has since been lost due to draining for agriculture and consequent land surface subsidence.[13]

Freshwater flows and flood control: The Sacramento River is now confined by flood-control levees that prevent land-water connectivity and divert flood flows to the Yolo Bypass. Dam building and water diversions within the watershed have likely caused an overall long-term decrease in flows and sediment delivery to the Delta. However, dams have increased summer flows in the Sacramento River, and other major tributaries, providing cold-water rearing habitat for salmonids that was formerly only present above dams.[14]

Sediment supply and transport: Flood-control levees along the lower reaches of many tributaries lock channels in place and shunt flood flows to the Central Delta. The Cosumnes River remains the only unregulated system feeding into the Delta and consequently flood flows regularly activate substantial portions of its floodplain.

Water diversions and flow regimes: Because of dams and water diversions, Delta inflows from the San Joaquin River have been greatly reduced. This reduction in flows, combined with water export operations, has caused unnatural flow reversals in the South Delta. Land surfaces are less subsided, but flows are largely constrained to leveed channels.

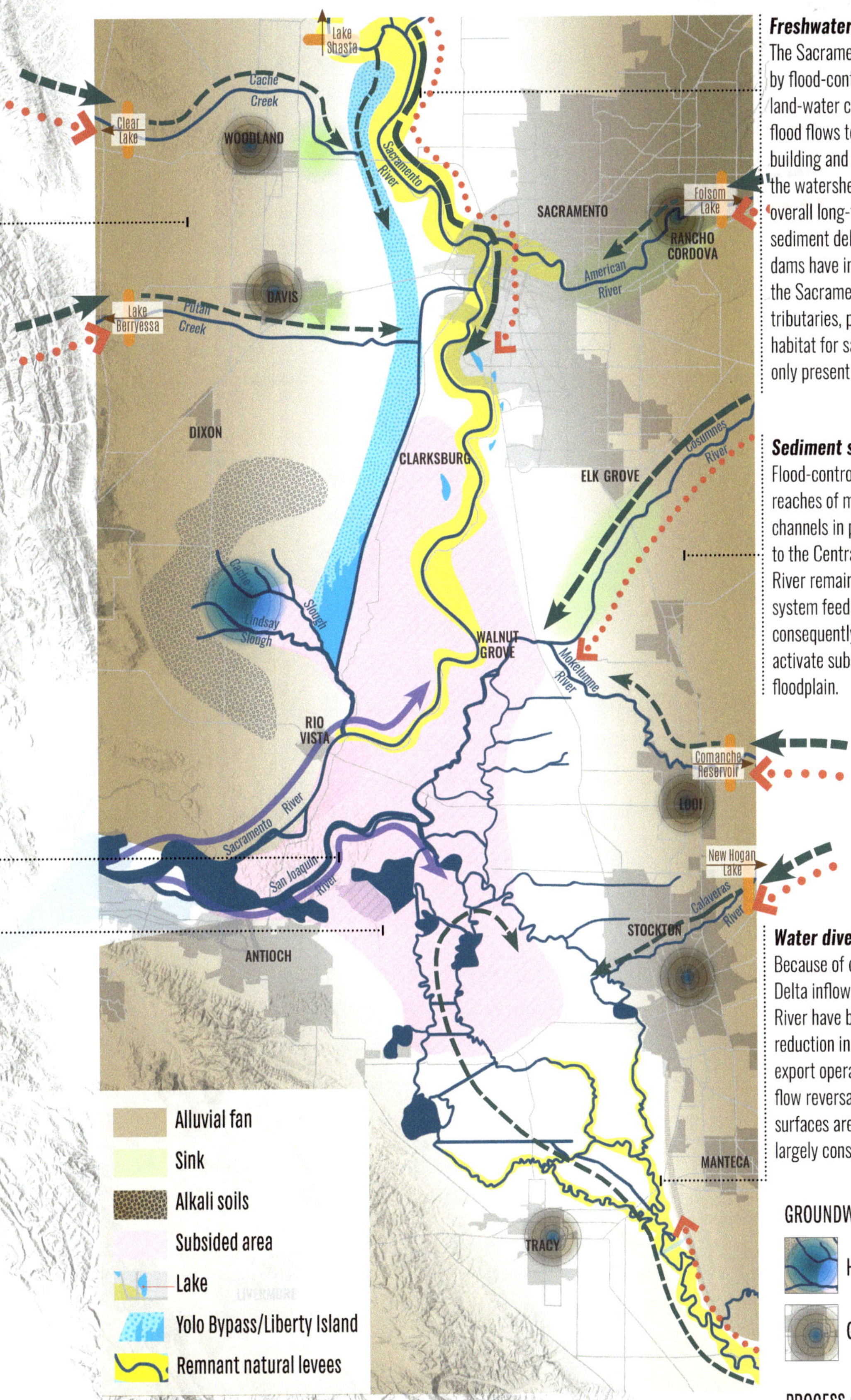

Lake Shasta
Cache Creek
Clear Lake
WOODLAND
Sacramento River
SACRAMENTO
Folsom Lake
RANCHO CORDOVA
American River
DAVIS
Lake Berryessa
Putah Creek
DIXON
CLARKSBURG
ELK GROVE
Cosumnes River
Cache Slough
Lindsay Slough
WALNUT GROVE
Mokelumne River
RIO VISTA
Comanche Reservoir
LODI
Sacramento River
San Joaquin River
New Hogan Lake
Calaveras River
STOCKTON
ANTIOCH
MANTECA
TRACY
Alluvial fan
Sink
Alkali soils
Subsided area
Lake
Yolo Bypass/Liberty Island
Remnant natural levees

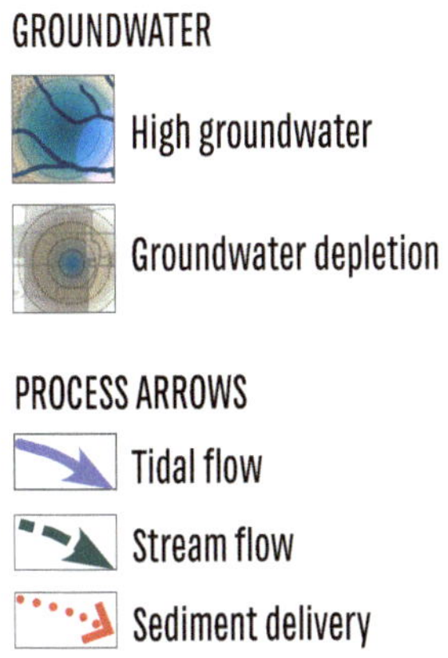

GROUNDWATER
High groundwater
Groundwater depletion
PROCESS ARROWS
Tidal flow
Stream flow
Sediment delivery

Historical Delta

NORTH

The Sacramento River, in the North Delta, was flanked by large natural levees (up to 20 ft high). At high flows, flood waters would spill over the natural levees and occupy low-lying flood basins, which would drain slowly, over months.

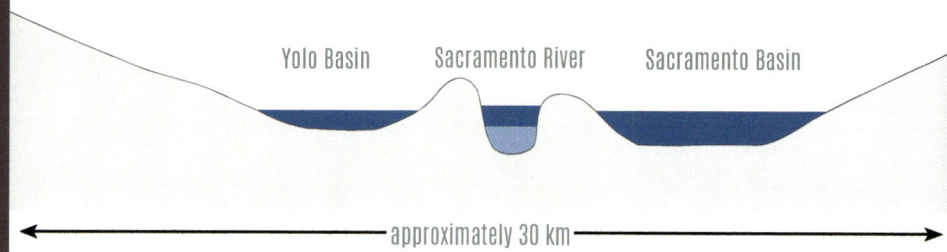

Yolo Basin   Sacramento River   Sacramento Basin

approximately 30 km

CENTRAL

The Central Delta, with its complex web of tidal sloughs and dead-end channels, would have been inundated perhaps daily and certainly monthly by the high tides, which would have covered the marsh plain for several hours at a time.

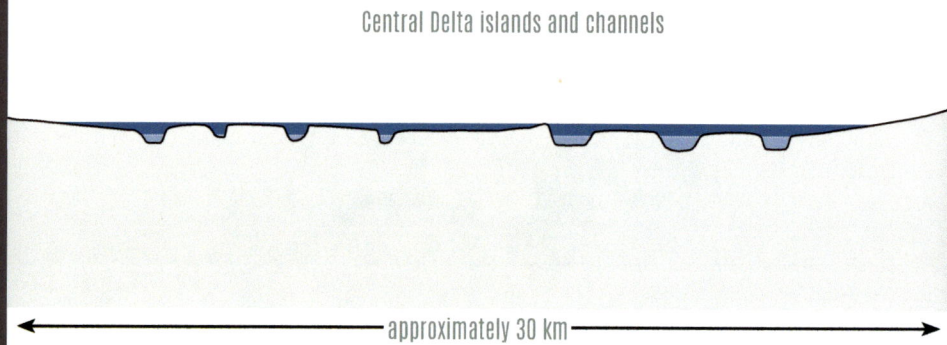

Central Delta islands and channels

approximately 30 km

SOUTH

The three main branches of the San Joaquin conveyed water at low flows and activated the many floodplains and side channels at high flows, particularly during the summer, often draining over the course of several months.[15]

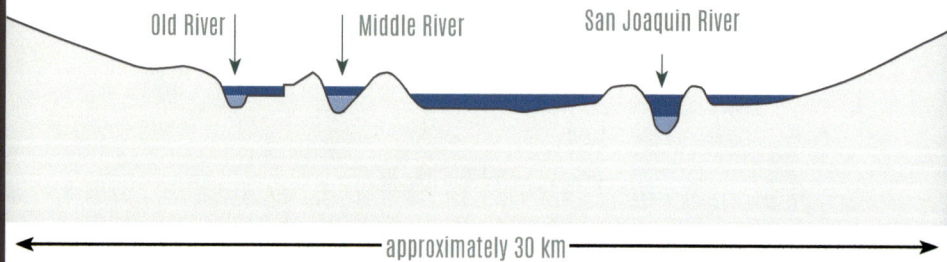

Old River   Middle River   San Joaquin River

approximately 30 km

High flow
Low flow

Modern Delta

NORTH

Flooding in the North Delta is generally confined to the flood-control levees along the river, and the Yolo Bypass, where inundation events lasting 7 days longer occur in approximately 25% of years.[16]

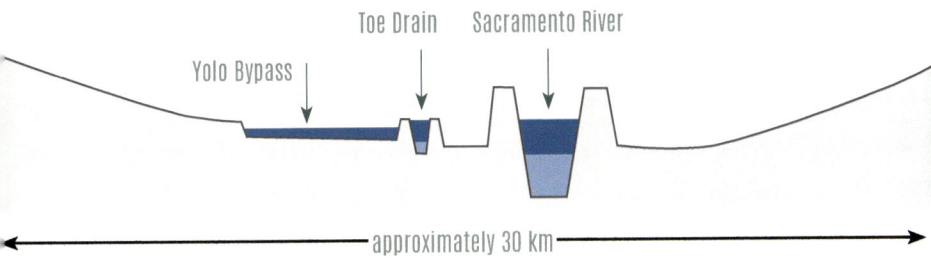

CENTRAL

The levees that protect the deeply subsided islands in the Central Delta were built to keep out flood waters. The land surface is often much lower than the water surface elevations.

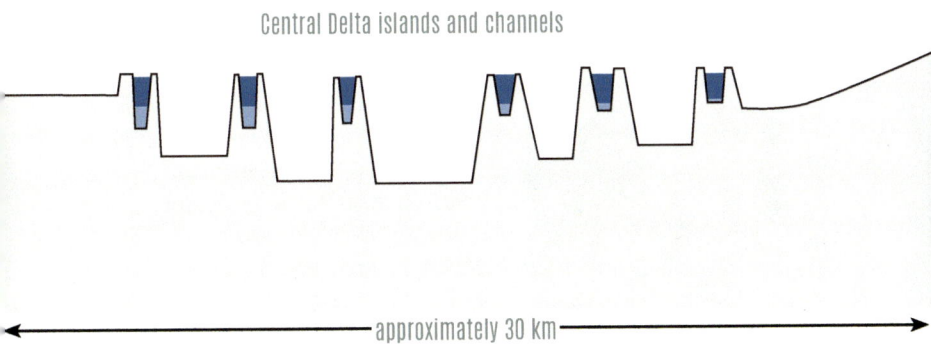

SOUTH

The three main branches of the San Joaquin are now confined by flood-control levees, leaving abandoned oxbows and side channels. Floodplain habitat has been greatly diminished.

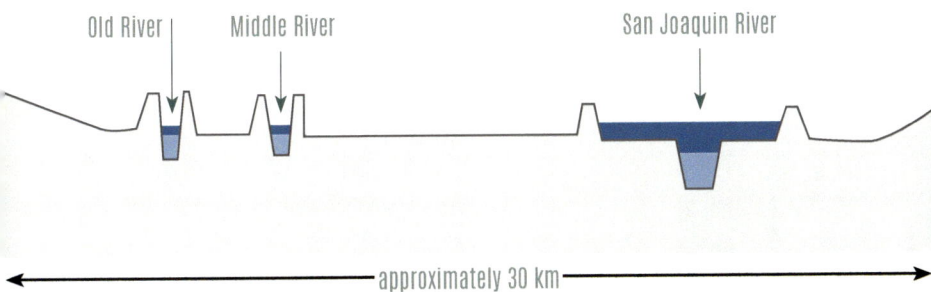

Yolo Bypass
Toe Drain
Sacramento River

approximately 30 km

Central Delta islands and channels

approximately 30 km

Old River
Middle River
San Joaquin River

approximately 30 km

High flow
Low flow

33

Ecological processes of the
Historical Delta

Ecological processes in the Delta were built upon, and interacted with, the template created by physical processes. The spreading and slowing of water across the Delta supported complex mosaics of wetlands and open water. Wetlands and shallow-water areas supported primary productivity that sustained robust food webs and large wildlife populations. Many producer groups contributed to the primary production in the Delta, including phytoplankton, surface algae, and marsh plants, supporting both direct consumption and detrital food-web pathways. Flows and flooding that connected open water, wetlands, and other floodplains allowed food and nutrients to be exchanged between them.

Connected habitat patches and flooding supported the movement of wildlife as they travelled to meet their daily needs, track variable resources, migrate, and disperse. The Delta's complex landscapes and connected habitats allowed wildlife to take advantage of variable conditions and diverse habitat types. For resident marsh and riparian forest species, large and continuous habitat patches were likely critical for maintaining large, genetically diverse populations. For migratory waterbirds and salmonids, the diversity of resource-rich wetlands and aquatic habitats allowed these highly mobile species to take advantage of different resources as conditions changed.

The physical and ecological processes in the Delta acted as selective forces to which species were constantly adapting. Tidal flows created predictable but highly variable patterns in hydrodynamics that influenced the vegetative structure of marshes and affected habitat conditions for aquatic species. Disturbance regimes created by flooding, fires, and other events, supported habitat complexity by creating space for colonization, a variety of successional stages, and redistribution of resources. This habitat complexity created different niches and sustained biodiversity in the Delta.

Floodplain inundation created connections between the estuary and West Delta tributaries, allowing for fish passage to the watersheds. *Activation of highly productive floodplain habitats* increased available food resources for rearing fish and allowed obligate floodplain spawners to reproduce. For waterbirds, variable flood depths supported a wide range of functional feeding groups (e.g., waders, divers, dabblers). Multiple flood basins, each with a somewhat different inundation regime, generated habitat redundancy and strengthened portfolio effects.

Alkali wetlands provided habitat for endemic species adapted to intermittent flooding. Intermittent floods lowered salinity of alkali soils, triggering seed germination and seedling establishment in many species. Diverse wetland types (e.g., intermittent, seasonal, and perennial; tidal and non-tidal; palustrine, riverine, and depressional) offered *complementary resources* to meet the various life-cycle requirements of many individual species. The different wetland types were also near one another and generally connected by terrestrial habitats, which facilitated the *migration of wildlife* between these wetlands.[17]

Wind-built *sand dunes* created *habitat complexity* within the tidal marshes of the Central Delta and provided high-tide refugia for marsh wildlife (and indigenous human settlements). The dunes served as habitat for a variety of endemic species, including Lange's metalmark butterfly, Antioch Dunes evening primrose, and western wallflower. Shifting sands *maintained early successional plant communities*.

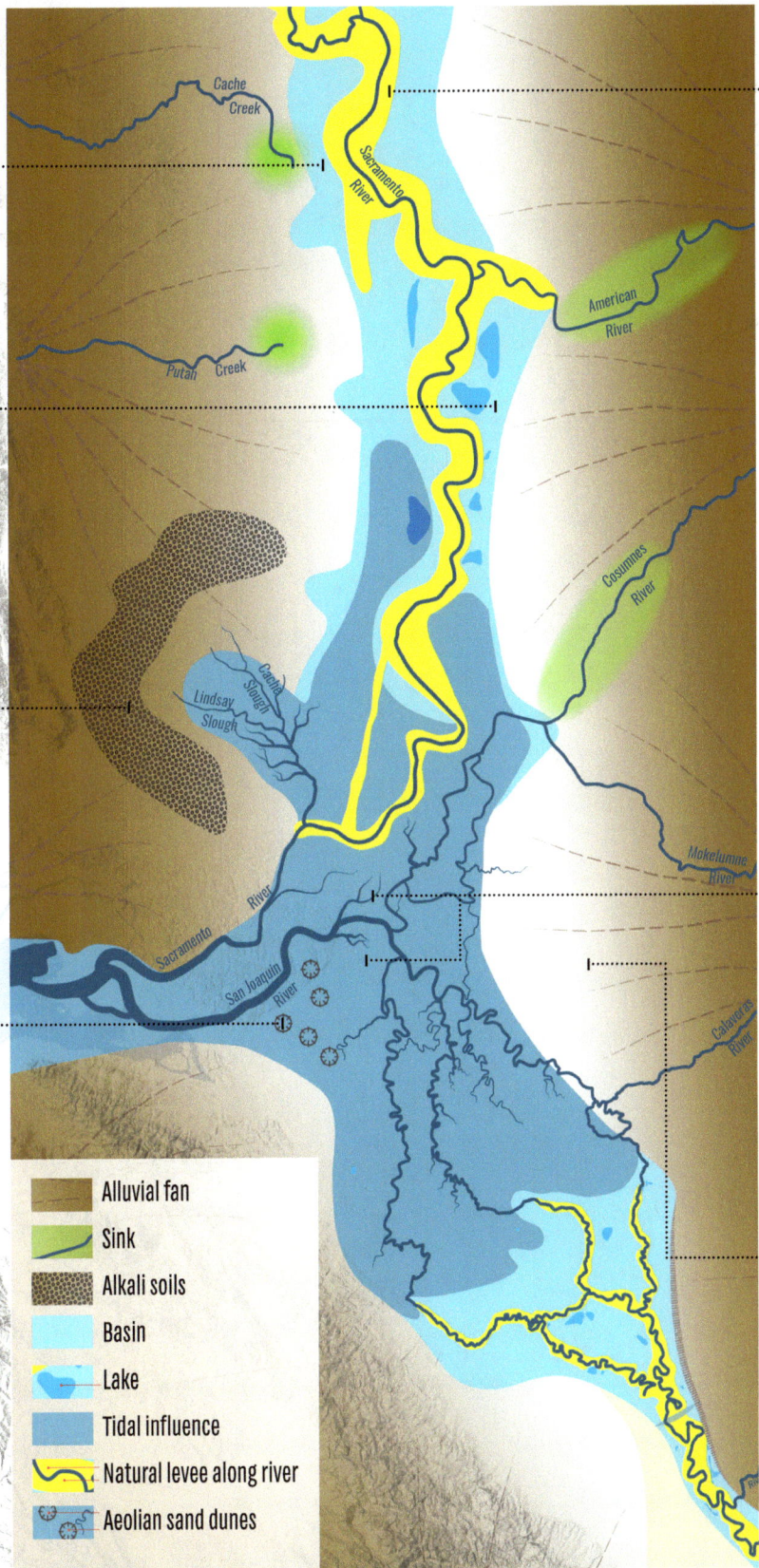

Cache Creek

Sacramento River

American River

Putah Creek

Cosumnes River

Lindsay Slough

Cache Slough

Mokelumne River

Sacramento River

San Joaquin River

Calaveras River

Legend

- Alluvial fan
- Sink
- Alkali soils
- Basin
- Lake
- Tidal influence
- Natural levee along river
- Aeolian sand dunes

Natural levees supported riparian forests, which shaded and contributed organic materials (including leaf litter and large woody debris) to the adjacent waterways, driving important aquatic **habitat complexity**. The raised land provided a **movement corridor** for terrestrial organisms and high-water refugia for wetland species. The physical structure and plant diversity of the forests provided food resources, cover, and nest sites for a variety of wildlife. Multiple woody riparian corridors created **redundancy** at the regional scale, safeguarding against the loss of whole populations.

Tidal inundation maintained a high groundwater table that allowed for the establishment of emergent wetlands and prevented the oxidation and decomposition of organic matter, which together created the buildup of peat over time. The result was equilibrium between land and water levels and **self-sustaining marsh elevations**. Tidal inundation also **connected marshes with aquatic habitats**, driving the flux of organisms, materials, and energy between the two environments. Tidal flows also **facilitated the daily and seasonal movements of many species**, including delta smelt. Multiple dendritic channel networks and large marsh patches contributed to structural and population redundancy. Waterbirds foraged in open water microhabitats of the marsh plain and nested in emergent vegetation.

Relatively well-drained **alluvial fans** supported an array of seasonal wetlands, grasslands, and oak woodlands and savannas along the Delta's periphery. These habitat types formed **broad transition zones** with the emergent wetlands of the Delta proper, which supported unique ecotone communities and allowed wildlife to move between the wetter and drier habitats (which is especially important for a variety of herpetofauna).[18]

35

Ecological processes of the
Modern Delta

Alteration of physical processes and other human activities have led to habitat loss, homogenization, and fragmentation in the modern Delta landscape. These changes, along with the introduction of numerous invasive species, have reshaped the biological structure of Delta wildlife communities and limited the expression of key ecological processes needed to support biodiversity. Habitat homogenization decreases diversity within populations and communities. Smaller, fragmented habitat patches are less able to support large, diverse populations, increasing the risk of extirpation, decreasing the likelihood of recolonization events, and reducing genetic diversity needed for future adaptation. Less connection between wetlands and other aquatic habitats impacts dispersal, gene flow, and other movement of biota, in addition to limiting the exchange of food, nutrients, and sediment.

The introduction of invasive and other non-native species has changed community and food-web structure through prey availability, competition, predation, and physical habitat changes that create novel niches. Changes in the physical and chemical environment (e.g., increased temperatures, changing hydrodynamics, changing suspended sediment concentration, and increased nutrients) can cause physiological stress and other impacts. Other human-related stressors (e.g., contaminants and feral cats) affect population size and selection pressures on existing populations.

On the other hand, the modern Delta also affords novel opportunities to support wildlife through targeted practices in agriculture, water management, and wildlife management.

Dams have *decreased connectivity* between the Delta and its watershed, preventing the upstream movement of anadromous fish to their spawning grounds. Dams have also reduced peak flows and *decreased the frequency, duration, and magnitude of floodplain inundation* and their associated ecological functions. *Reduced sediment supply* and *reduction in peak flows*, along with artificial levees, have also reduced sediment transport and deposition, resulting in reduced floodplain habitat heterogeneity. Rapid recession of floodplain flows and barriers to upstream movement on the Yolo Bypass increase the risk of stranding fish.

The *reduced extent and diversity of wetland* types around the Delta's periphery likely contributes to *decreased stability of regional wildlife populations*, because wetland types (e.g. perennial and intermittent) with travel corridors between them offer complimentary resources for wildlife over time and space.[19]

Sparsely vegetated *artificial levees* are subject to *severe edge effects* and serve as *poor movement corridors* without functional connections to the aquatic environment.

Remnant sand dunes: Exotic grasses have contributed to the *increased stabilization of remnant dunes* and inhibited the growth and recruitment of native early successional plant species (including buckwheat, the host for the endangered Lange's metalmark butterfly). In an effort to restore shifting sandy substrate, dredged Bay sediment is occasionally applied directly to the remnant dune system.[20]

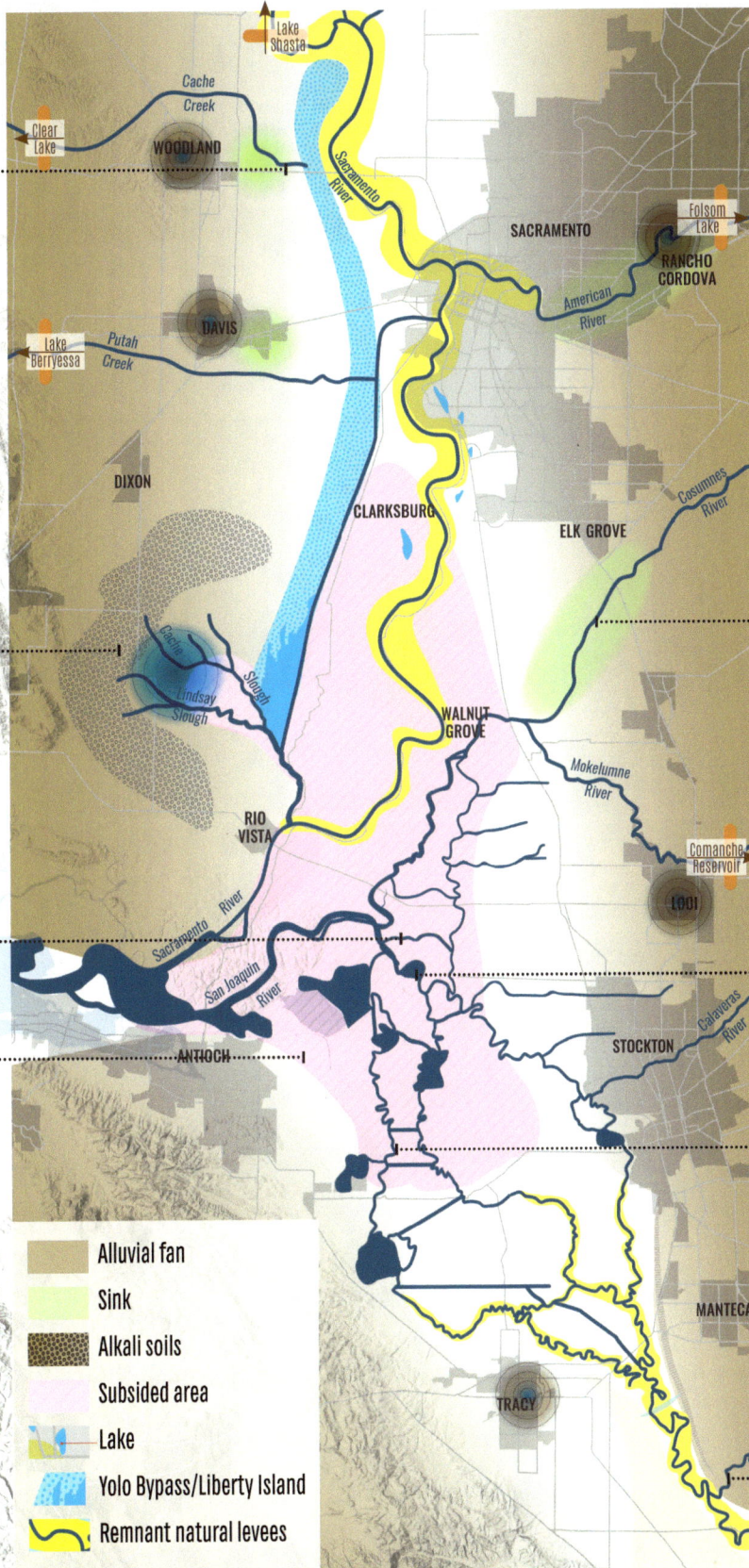

Lake Shasta

Cache Creek

Clear Lake

WOODLAND

Sacramento River

SACRAMENTO

Folsom Lake

RANCHO CORDOVA

American River

DAVIS

Lake Berryessa

Putah Creek

DIXON

CLARKSBURG

ELK GROVE

Cosumnes River

Cache Slough

Lindsay Slough

WALNUT GROVE

Mokelumne River

RIO VISTA

Comanche Reservoir

LODI

Sacramento River

San Joaquin River

STOCKTON

Calaveras River

ANTIOCH

MANTECA

TRACY

The unregulated **Cosumnes River** provides an *intermittent movement corridor* for fish and sediment, except where ground water pumping dries lower reaches for much of the year. New areas of woody riparian forest provide extensive habitat for associated wildlife.

Extreme **marsh loss** has **reduced marsh-derived primary productivity** as well as the retention of nutrients in the Delta. Remaining marsh fragments are small and isolated, which means they support smaller wildlife populations with less gene flow.

Altered fluvial flows negatively affect native fish habitat, movement, and navigation. Effects are particularly severe in the South Delta.

Extensive wide **woody riparian** areas along the lower Stanislaus River provide habitat for endangered riparian mammals. However, during flood events, artificial levees increase the height of the water, causing severe mortality in the already-diminished populations.[21] The **lack of redundant habitat patches** and wildlife populations increases the risk of extinction.

Legend

- Alluvial fan
- Sink
- Alkali soils
- Subsided area
- Lake
- Yolo Bypass/Liberty Island
- Remnant natural levees

GUIDANCE FOR IMPLEMENTING PROCESS-BASED STRATEGIES IN THE DELTA

04

Photograph by Shira Bezalel (SFEI-ASC)

Restoration and management actions that incorporate naturalistic physical processes will be essential to conferring desired ecological resilience to the Delta, particularly in light of sea-level rise and other climate-change impacts. In this section, we describe the processes that could be established to create and maintain an ecologically healthy system in the Delta. We draw from the conceptual models in Chapter 3 to provide guidance for how key physical processes should work, and offer landscape specifications for related habitat types. These recommendations integrate research done by other experts with our understanding of processes in the Delta, based on our previous studies of the historical conditions and landscape change. The strategies and specifications presented here are meant to provide guidance at both the landscape and project scales.

PROCESS-BASED STRATEGIES

Restore **tidal zone** processes	Re-establish tidal marsh processes in areas at **intertidal elevations** Re-establish marsh processes in **subsided areas** Re-establish tidal zone processes in **channels and flooded islands**
Restore **tidal-fluvial transition zone** processes	Re-establish **connections between streams and tidal floodplains**
Restore **fluvial zone** processes	Re-establish **fluvial processes** along streams
Restore **terrestrial and transition zone** processes	Re-establish **tidal-terrestrial transition zone processes** Re-establish **connected terrestrial habitats** around the periphery of the Delta
Integrate ecological processes with **human land uses**	Expand **wildlife-friendly agriculture** Integrate **ecological functions** into urban areas

Why process-based restoration?

Many management approaches have been tried to improve degraded aquatic ecosystems. Conventional restoration often focuses on creating specific habitat characteristics across acreage quotas to meet habitat needs for particular species.[1] This method becomes problematic when these static habitat configurations cannot respond to disturbance events, such as floods or droughts, or cannot adequately sustain themselves over time—essentially, when these habitat patches are not resilient. For example, if a tidal marsh is restored but not fully connected to the tides or to a sufficient sediment supply, the marsh will require ongoing investment in engineering for long-term viability as sea level rises.

Process-based restoration aims to address the primary causes of ecosystem degradation[2] and has been shown to be more effective over the long term than conventional approaches. For example, addressing sediment source, transport, and delivery in a sediment-starved reach of a river may be more successful at creating and sustaining suitable salmon habitat than creating in-channel structures.[3] If done successfully, restoring physical, chemical, and biological processes that create and maintain habitat requirements, rather than creating just the habitat requirements, can prove more sustainable and beneficial for ecological functioning, and potentially require less management in the long term.[4]

Because these processes take place over long time scales (years to decades, and even longer), management of both the physical site as well as human expectations around project outcomes will be critical for success. Many solutions that use natural processes will still require and benefit from active management, especially in a system as altered as the Delta. Engineering solutions, such as tide gates and water control structures, and land management solutions, such as acquisition, easements, and wildlife-friendly farming, are tools to this end.

The long-term aim of process-based restoration is to create dynamic, resilient ecosystems that provide desired ecological functions over long time scales with minimal intervention, rather than static habitat patches. Restoration of critical processes (including halting anthropogenic disruption of these processes)[5] enables the system to more resiliently adapt over time—recovering after disturbance and continuing to function as desired as baselines shift. Both types of resilience will become more important as climate change accelerates.[6] This approach to restoration requires large spatial scales, long time frames, and the ability to coordinate across the complexity of the Delta landscape, including a multitude of landowners, regulations, and land uses.

Process-based strategies

We focus on the large-scale earth-surface processes that have been interrupted in the Delta and that can create the most positive change upon being re-established or emulated. Namely, these are the semi-diurnal tides, their interaction with bathymetry/topography, and the movement of water and sediment between terrestrial and aquatic systems. From these major processes, we developed a set of restoration strategies, which could be used to re-establish or enhance the physical and biological processes that have been interrupted and diminished in the Delta. The five major strategy groups, with sub-strategies, and the habitat types they usually apply to are shown in the table below.

Some of the historical habitat types, particularly tidal and non-tidal marshes, span more than one "zone" in the Delta. Multiple strategies are appropriate for restoration and conservation of these habitats, and supporting the varied processes involved leads to diverse and dynamic systems. In this chapter we primarily discuss landscape specifications for tidal marsh on pp. 44-47 and non-tidal freshwater marshes on p. 60. Novel analogues can mimic historical habitat types and the processes that support them, but may not fit as easily into these "zone" classifications depending on the specific site history and management.

Nine strategies for process-based restoration in the Delta, organized into groups by broad physical zones. Habitat types associated with each strategy group are also listed for reference.

Strategy Group	Strategies	Associated habitat types
Restore tidal zone processes	Re-establish tidal marsh processes in areas at intertidal elevations Re-establish marsh processes in subsided areas Re-establish tidal zone processes in channels and flooded islands	**Historical:** tidal freshwater emergent wetland, tidal channels, submerged and floating aquatic vegetation (SAV/FAV) **Novel:** flooded islands, managed wetlands, non-native SAV/FAV
Restore tidal-fluvial transition zone processes	Re-establish connections between streams and tidal floodplains	**Historical:** tidal freshwater emergent wetland, non-tidal freshwater emergent wetland, freshwater pond or lake, tidal and fluvial channels, willow thicket **Novel:** managed wetlands, flooded islands
Restore fluvial zone processes	Re-establish fluvial processes along streams	**Historical:** fluvial channels, non-tidal freshwater emergent wetland (including flood basins and floodplains), valley foothill riparian, willow riparian scrub or shrub, willow thicket **Novel:** willow scrub on artificial levees, flood bypasses, managed floodplains
Restore terrestrial and transition zone processes	Re-establish tidal-terrestrial transition zone processes Re-establish connected terrestrial habitats around the periphery of the Delta	**Historical:** tidal freshwater emergent wetland (high marsh ecotone), non-tidal freshwater emergent wetland, wet meadow or seasonal wetland, vernal pool complex, alkali seasonal wetland complex, stabilized interior dune vegetation, grassland, oak woodland or savannah **Novel:** willow scrub on artificial levees, managed wetland, annual grasslands
Integrate ecological processes with human land uses	Expand wildlife-friendly agriculture Integrate ecological functions into urban areas	**Novel:** wildlife-friendly agriculture, green infrastructure

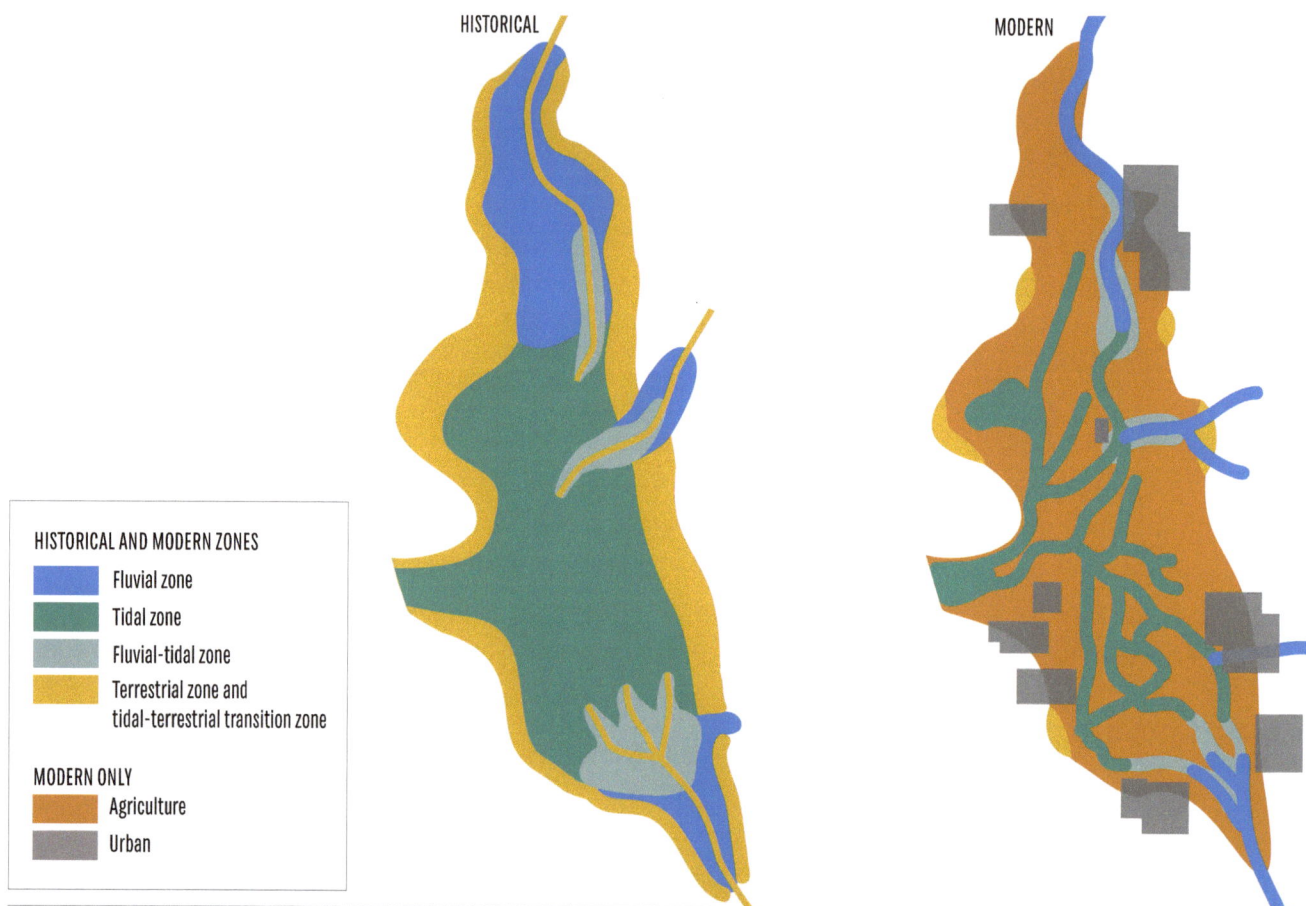

HISTORICAL   MODERN

HISTORICAL AND MODERN ZONES
- Fluvial zone
- Tidal zone
- Fluvial-tidal zone
- Terrestrial zone and tidal-terrestrial transition zone

MODERN ONLY
- Agriculture
- Urban

These strategies will fit into the current and future Delta landscapes in ways that may not match their historical locations and configurations. In these conceptual maps of the historical and modern Delta, we show examples of where these process zones occur in the historical and modern Delta. Places where these strategies may be best employed in the future Delta may not match either the past or present distributions. In describing each strategy, we identify some of the factors important to identifying opportunity areas for restoring appropriate processes. For example, while historically the tidal zone encompassed most of the core of the Delta, today the tidal zone is restricted to the channels in tidal influence, as well as some fringing tidal marsh. The maps cannot show all the detail of the real Delta. For example, every tributary to the Delta has a fluvial-tidal zone—not just the three pictured here.

In the next several pages, we describe the dominant processes that occur within each of these process-based strategy groups. For each strategy, we describe specific physical processes that are recommended to implement the strategy in question (shown in orange), and guidelines pertaining to the configuration and scale of landscape elements (shown in green). The landscape configuration and scale guidelines are drawn from the landscape ecology metrics produced in our earlier *Delta Transformed* report.[1]

Supporting particular wildlife groups and other ecological functions in the Delta will require layering multiple strategies in varying configurations, across temporal and spatial scales. The guidelines presented for these strategies should be incorporated with other key issues in on-the-ground resource management, such as phasing of projects across time, land-ownership, permitting, engineering requirements, and monitoring. In the subsequent chapter (Chapter 5), we describe how these strategies can fit together in the landscape to maximize desired ecological functions in the future Delta.

Photograph by Shira Bezalel (SFEI-ASC)

Re-establish tidal marsh processes in areas at **intertidal elevations**

SUPPORTED FUNCTIONS

marsh wildlife
(see pp. 88-91)

edge wildlife
(see pp. 100-103)

waterbirds
(see pp. 96-99)

fish
(see pp. 84-97)

biodiversity
(see pp. 106-107)

productivity
(see pp. 104-105)

Large swaths of land in the Delta currently are situated at intertidal elevations but are separated from the tides by levees and other human infrastructure. These areas have the greatest potential to support tidal marshes with minimal management intervention now and into the future because, if connected to tidal action, they would be inundated at a depth and frequency that is appropriate for the establishment and persistence of emergent marsh vegetation. The ebb and flow of the tides across intertidal areas would allow for processes—like sediment deposition, scour, and flooding—that are needed to create channel networks, ponded areas, natural levees, and other important marsh features.

The areas at intertidal elevation are not static. With continued sea-level rise (SLR), these areas will shift. This process generates opportunities to restore tidal marshes in new (landward) areas in the future, but also increases the urgency to restore the areas that are currently intertidal, while their elevation is still favorable. More research is needed to understand the trade-offs that accompany large tidal marsh restoration. One concern is that an increase in the area of tidal marsh could alter tidal range in other parts of the Delta, with cascading effects that are difficult to predict.

Specific tactics for carrying out this strategy include: connecting lands in intertidal areas to tidal action by removing or breaching levees; removing other barriers to the exchange of water and sediment across marsh surfaces; and preventing the erosion of marsh edges using wind/wave energy breaks and other shoreline stabilization structures (both living and non-living). Many of these tactics will require active management using tide gates and water-control structures.

Image courtesy Google

Delta Meadows
(some existing tidal marsh)

McCormack Williamson Tract

current MHHW + 6 ft (1.8
(generalized)

area at intertidal elevation
(now and into future)

current MLLW (generalized)

Intertidal elevations, McCormack-Williamson Tract: The outlined portion of this landscape either currently sits at an intertidal elevation or will in the future with approximately 6 ft (1.8 m) of SLR. Since a significant portion of the McCormack Williamson Tract falls in this elevation range, plans are underway to re-establish tidal processes over much of the island.

North-central Delta: Largest contiguous area of land at intertidal elevation in the North Delta. Opportunity will be lost with just a few feet of SLR. There are currently no plans to restore natural habitat types in the area, nor are there currently any protected lands which could be considered for restoration. It is characterized by high-value agricultural land, including annual row crops, vineyards, and orchards.

Northwest Delta: Large, wide areas at intertidal elevation with good connection to natural landward habitats (seasonal wetlands) and to Suisun Marsh. Opportunity to enhance connectivity between existing large marsh patches (those at Liberty Island and Lindsey Slough). Some protected land. Restoration planning underway.

Landscape considerations
to re-establish tidal marsh processes in areas at intertidal elevations

This map highlights areas that are currently at intertidal elevation or will be in the relatively near future (with 3 and 6 ft [0.9 and 1.8 m] SLR). It also shows urbanized areas, which generally should not be considered for this strategy, and existing marshes, which helps identify regions that are lacking large marshes and where stepping-stone marshes might be most beneficial. Although difficult to calculate precisely, our map shows approximately 35,000 ha of land currently at intertidal elevation (of which 32,000 ha are not urbanized). An additional 40,000 ha of non-urbanized land would be situated at intertidal elevations with 6 ft [1.8 m] of SLR, which would bring the total area of non-urban land situated at the right elevation for this strategy to 72,000 ha. This area is 23 times larger than the current 3,000 ha of tidal marsh[8] and close to 50% of the historical extent.[9]

Northeast Delta: Large areas at intertidal elevations contiguous with existing natural habitat types (wetland, riparian, and terrestrial habitats associated with Stone Lakes, Delta Meadows, and Cosumnes River). Restoration already planned for McCormack-Williamson Tract. Opportunities for connection to unregulated Cosumnes River and its floodplains.

Southwest Delta: Opportunity to restore marshes at the low-salinity transition between Suisun Bay and the rest of the Delta. Restoration planned at Dutch Slough. Potential to recreate corridor between existing marshes of Sherman Island, Sand Mound Slough, and other small remnant stepping stones. Some protected land, but generally located below intertidal elevations. Some agricultural areas and natural habitats on landward margin, but many areas constrained by urban development.

Eastern margin: Continuous band of intertidal area more than 3 km wide (with 6 ft [1.8 m] SLR). Connected to remnant dead-end sloughs (see pp. 52-55) with stepping stone remnant marshes reaching into the Central Delta. Landward edge bounded by Interstate 5 (I-5), but otherwise agricultural. Some tracts partially subsided, so new infrastructure and/or reverse subsidence would be required to breach portions currently at the correct elevation (see pp. 48-51). Limited protected land for restoration.

South Delta: Largest contiguous area of land at intertidal elevation in the Delta. Opportunities in conjunction with flood-protection planning to re-establish connections to San Joaquin River floodplains, sediment supply, and riparian habitats, which could promote resilience of tidal marshes to future SLR (see pp. 56-59). Almost no existing large marshes in the region. Currently serves as highly productive agriculture. No protected/public land.

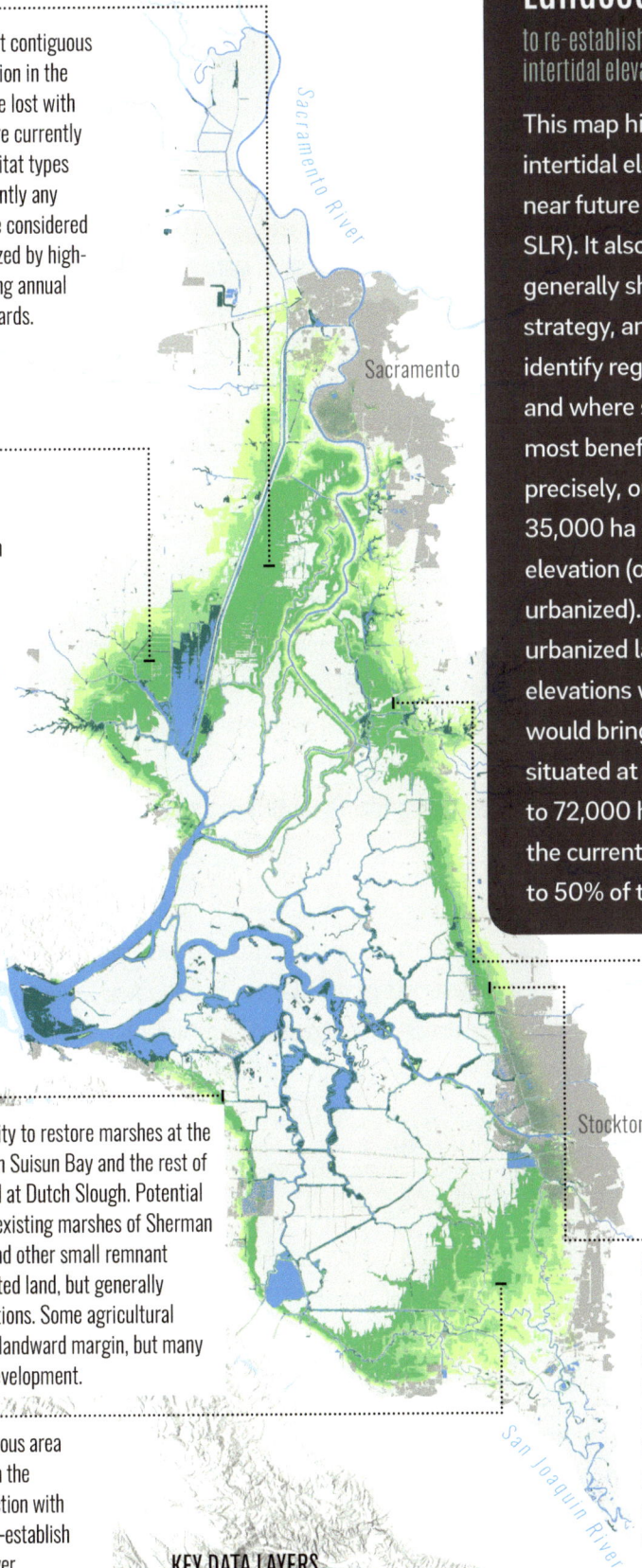

Sacramento River

Sacramento

Suisun Bay

Stockton

San Joaquin River

KEY DATA LAYERS

Intertidal elevation[10]
- currently intertidal
- currently intertidal +3 ft (0.9 m)
- currently intertidal +6 ft (1.8 m)

Existing marshes

Urbanized areas

N

10 miles

SFEI AQUATIC SCIENCE CENTER

PHYSICAL PROCESS GUIDELINES

1 **Tidal marshes should experience full tidal action**

Tidal flows should be sufficient to drive the flux of materials into and out of marshes. In particular, tidal flows should drive regular inundation of the marsh plain, which provides direct access to foraging by aquatic organisms, enhances the export of productivity from the marsh plain into the adjacent aquatic environment, plays a role in maintaining local water temperature gradients, and promotes marsh accretion (see Guideline #3 below). Tidal flows should also be sufficient to drive the formation and maintenance of dendritic channel networks, which increase habitat complexity and are critical to the use of marshes by many species.

2 **Tidal marshes should have complex and variable patterns of tidal inundation**

Marshes naturally exhibit complex patterns of inundation and drainage driven by the feedbacks between spring-neap variability in tidal range and microtopographic features. Lower high tides inundate the marsh plain from the tips of interior blind channels, while higher high tides inundate the marsh plain over small natural levees around its perimeter.[11] Ecosystem engineers, such as beaver and waterfowl, also alter the marsh surface and add to its spatial heterogeneity.[12] Variable inundation patterns drive fine-scale heterogeneity in environmental conditions (such as soil moisture and chemistry) and create different microhabitats for a range of plants and animals.

3 **Tidal marshes should maintain processes that allow them to keep their extent over time**

For more than 6,000 years, as sea level in the estuary steadily rose, the Delta's marshes maintained land-surface elevations slightly above local mean sea level. Multiple interrelated processes contributed to this homeostasis and allowed for the continuous existence of marshes over time, but of particular importance is the vertical accumulation of organic plant matter. Frequent inundation (tidal or fluvial) is critical to the accumulation of organic matter, since it helps maintain high water table levels that prevent the oxidation and decomposition of peat.[13] Inundation also contributes to marsh accretion through 1) the delivery of suspended inorganic sediment, which contributes to peat formation, and 2) the delivery of nutrients, which promote plant growth and the accumulation of organic material.[14]

LANDSCAPE CONFIGURATION & SCALE GUIDELINES

4 **Tidal marshes should be as large as possible**

Though small marshes have some value, marshes should be as large as possible since the functions they support increase with size. For example, marshes as small as 1 ha can support some California Black Rails, but the density of rails is maximized once marshes reach approximately 100 ha in size. Blind channel length also increases disproportionately with marsh island area;[15] marshes larger than most that exist today are likely needed to maintain long, multi-order channel networks (see pp. 52-55).

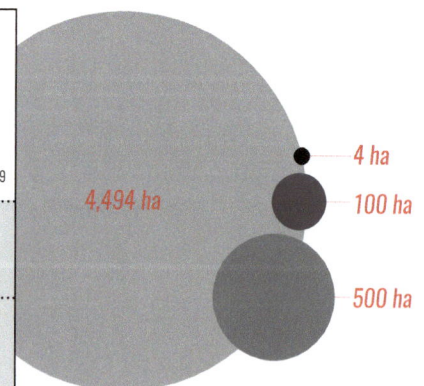

<1 ha = l marsh patch size for Tricolored Blackbird nesting[16]

1 ha = minimum marsh patch size for California Black Rail occupancy[17]

100 ha = minimum marsh patch size for maximum Black Rail density[18]

500 ha = approximate marsh area for a full channel network (based on historical landscape)[19]

4,494 ha = average **historical** patch size (SD = 17,956)[20]

4 ha = average **modern** patch size (SD = 24)[21]

110,527 ha = maximum **historical** patch size[22]

749 ha = maximum **modern** patch size[23]

Reference values

4,494 ha

4 ha

100 ha

500 ha

5 **Distance between tidal marshes should be minimized**

Restoration plans should aim to decrease the nearest neighbor distance of Delta marshes and increase the proportion of marshes that occur in close proximity to large marshes. Marsh nearest neighbor distances should be informed by factors like animal dispersal distances. For example, because outmigrating juvenile salmon travel during the night and hold in low-velocity refugia habitats like marsh channels during the day,[24] they may benefit from gaps between marshes that are less than the distances they generally travel over a 24 hour period. Though historically the maximum distance between marshes was much less than this distance, today even the mean distance between marshes exceeds the mean distance smolts generally travel in a day.

0.2 km = median natal Song Sparrow dispersal distance (San Pablo Bay)[25]

5 km = mean Black Rail dispersal distance[26]

15 km = mean salmon smolt daily migration distance[27]

0.3 km = mean **historical** distance from one marsh to a sizeable (100 ha) marsh (SD = 0.4)[28]

19.2 km = mean **modern** distance from one marsh to a sizeable (100 ha) marsh (SD = 11.1)[29]

1.6 km = maximum **historical** distance from one marsh to a sizeable (100 ha) marsh[30]

61.4 km = maximum **modern** distance from one marsh to a sizeable (100 ha) marsh[31]

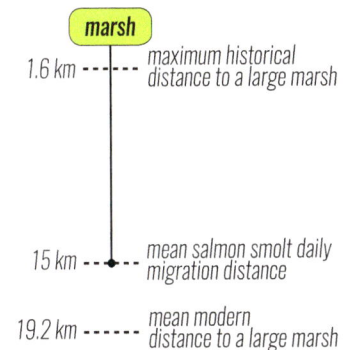

marsh

1.6 km ------ *maximum historical distance to a large marsh*

15 km ------ *mean salmon smolt daily migration distance*

19.2 km ------ *mean modern distance to a large marsh*

6 **The ratio of core to edge habitat should be maximized**

Marsh patches should have more core habitat than edge habitat (excluding "interior" edges created by channel networks). Core areas experience distinct abiotic conditions, are less accessible to many predators of marsh wildlife, and are more buffered from human disturbance in the modern landscape. We would expect, for example, Black Rail presence to be more likely in patches with high core to edge ratios than those with low ratios.[32]

13.1 = **historical** marsh core:edge area ratio[33]

0.2 = **modern** marsh core:edge area ratio[34]

7 **The ratio of marsh to open water should increase**

Individual restoration projects should increase the landscape's marsh to open water ratio. Increasing the ratio would be expected to increase the availability of marsh-derived primary productivity to the aquatic food web. This is important since most large estuaries depend on detrital pathways to fuel the food web.[35] Research suggests that pools of particulate organic carbon (POC) in the aquatic environment will only reflect marsh inputs when total marsh area exceeds total open water area.[36]

1.0 = approximate minimum marsh : open water area ratio for marsh-derived carbon to be reflected in open water POC pools[37]

11.8 = **historical** marsh:open water area ratio[38]

0.2 = **modern** marsh:open water area ratio[39]

8 **Maximize tidal marsh-water edge length through the development of interior channel networks**

Adjacency between marshes and open water habitats is required for many aquatic organisms to utilize and benefit from marshes. Increasing the length of adjacency through the fragmentation of existing marshes would be counterproductive (see Guideline #6 above). Adjacency should instead be increased by developing channel networks embedded within marshes (see pp. 52-55).

Re-establish marsh processes in **subsided areas**

SUPPORTED FUNCTIONS

marsh wildlife
(see pp. 88-91)

edge wildlife
(see pp. 100-103)

waterbirds
(see pp. 96-99)

fish
(see pp. 84-87)

biodiversity
(see pp. 106-107)

productivity
(see pp. 104-105)

Since reclamation, the Delta's peat soils, particularly those of the Central Delta, have supported a highly productive agricultural industry. However, due to the oxidation and compaction of peat associated with agricultural practices, this land use has resulted in as much as 8 m (26 ft) of subsidence, a loss of more that 3,000 years worth of accretion.[40] The most severe subsidence has occurred in the Central Delta, where peats were the deepest, while surface elevations at the tidal margins have subsided less. This history has created a subtidal "bowl" that must be protected by levees to support current land uses. The risks to the current configuration are great: continued subsidence, earthquakes, and sea-level rise threaten to cause failures in the brittle system of levees.

In light of this situation, strategies to re-establish tidal processes in key subsided areas will require us to rebuild lost elevation ("reverse subsidence"). Doing so would require careful management of accretion and depositional processes over many decades. Specifically, there is some potential to reverse subsidence in the Delta by establishing managed wetlands with a hydrologic regime designed to maximize the accretion of biomass over long time frames. Known colloquially as "tule farming," this process has been tested at two sites in the Delta, where maximum land-surface elevation gains of 7–9 cm per year have been achieved.[41] Though the ultimate goal of tule farming might be to restore tidal marshes, the subsided managed wetlands would not be tidal during the interim period (though they could still provide a subset of the desirable ecosystem functions provided by tidal marshes). Elevations could also be built through the direct placement of sediment[42] or through related tactics like warping, in which sediment deposition rates are maximized by repeatedly flooding parcels long enough for suspended sediment to fall from the water column.[43]

Care must be taken when choosing locations to attempt reverse subsidence at a large scale, since there is a risk that levees could fail prior to the recovery of target elevations. Simple models can help prioritize investments in reverse subsidence using variables such as the degree of subsidence, annual probability of levee failure, and the amount of agricultural revenue that would be lost with conversion to wetlands have been developed.[44] Ultimately, reverse subsidence efforts might only be practical in minimally-subsided areas.

Image courtesy Google

Reverse subsidence project, Twitchell Island: Reverse subsidence methods are currently being tested at Twitchell Island, where both rice fields and managed wetlands have been established to test how they affect land-surface elevations over time.

rice farming subsidence reversal demonstration project

managed wetland subsidence reversal demonstration project

Northeast Delta: Multiple tracts (including the McCormack-Williamson Tract) that are only minimally subsided (25–50 year outlook) and are contiguous with areas currently at intertidal elevation.

Cache Slough Complex: Sizeable areas at the bottom of Egbert, Little Egbert, and Hastings tracts are minimally subsided, contiguous with lands currently at intertidal elevations, and proximal to relatively large existing tidal marshes, but are not in public ownership. The low end of the Cache-Haas area and most of Liberty Island are both minimally subsided and publicly owned (currently, these areas are permanently flooded). Restoration planning is underway, making these areas of high potential for reverse subsidence strategies.

Permanently flooded islands: Most of the Delta's permanently flooded islands—including Franks Tract, Liberty Island, Sherman Lake, and Big Break—are all relatively un-subsided. Franks Tract, in particular, is much less subsided than the surrounding islands, which might make it a good area to prioritize for reverse subsidence strategies like the direct placement of sediment.

Landscape considerations
to re-establish marsh processes in subsided areas

This map shows subsided areas in the Delta, color-coded based on the approximate amount of time it would take to recover intertidal elevations through tule farming, given the simplifying assumptions that 1) the land surface elevation can be increased at a rate of 5 cm/year and 2) sea levels do not continue to rise.[45] It also shows existing marshes and areas that are currently at intertidal elevation, which are often contiguous with minimally subsided areas, that should be considered for reverse subsidence efforts. Urbanized areas, which shouldn't be considered for this strategy, are also shown. Approximately 50,000 ha of non-urban leveed areas are currently subsided by less than 1.9 m (which would take approximately 50 years to recover, given the assumptions stated above).

East Delta: Multiple tracts are minimally subsided (25–50 year outlook), but, if restored, would have minimal transition zone and accommodation space (also see pp. 66-69). No existing public/protected land.

South Delta: Numerous large islands are only minimally subsided. These areas are also contiguous with wide areas of land that are currently at intertidal elevations, which enhances the potential for re-establishing tidal zone processes. Organic matter accretion in this area could potentially be supplemented with the deposition of inorganic sediments from the San Joaquin River (see pp. 56-59). However, the area is characterized by productive agriculture and there is currently no public/protected land.

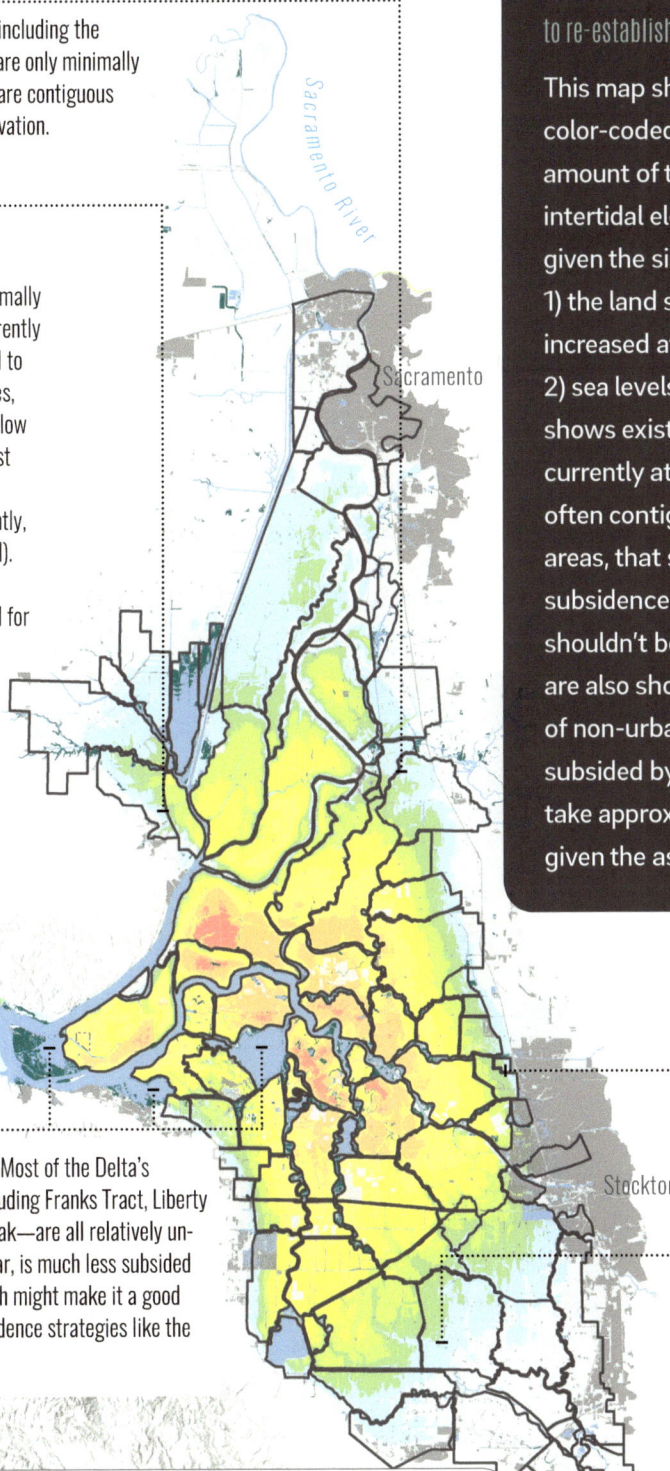

KEY DATA LAYERS

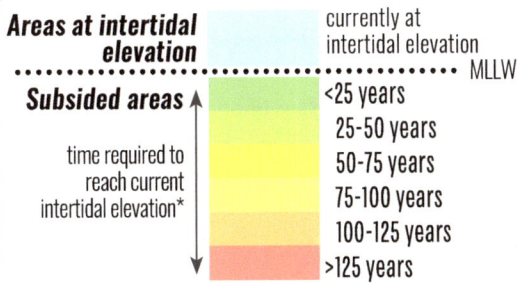

Areas at intertidal elevation

currently at intertidal elevation — MLLW

Subsided areas

time required to reach current intertidal elevation*

- <25 years
- 25-50 years
- 50-75 years
- 75-100 years
- 100-125 years
- >125 years

* values assume constant elevation gains through tule farming of 5 cm per year and no sea-level rise

Existing marshes

Urbanized areas

Islands and tracts

dark grey outlines on map denote boundaries of individual islands and tracts

N

10 miles

Sacramento River

Sacramento

Suisun Bay

Stockton

San Joaquin River

PHYSICAL PROCESS GUIDELINES

1 **Halt or limit ongoing subsidence by increasing groundwater levels**

The height of the water table determines the depth to which the organic carbon in peat can be oxidized (converted to carbon dioxide by microorganisms) and therefore exerts a strong control on the rate of ongoing subsidence. Since oxidative peat loss will continue as long as the water table is held below the land surface, groundwater recharge (especially through regular tidal inundation) should be a priority wherever reverse subsidence is being considered.[46]

2 **Maximize the vertical accumulation of organic matter by optimizing managed wetland hydrology**

Through pilot reverse subsidence projects, it has been determined that site hydrology drives critical differences in the biological and biogeochemical processes responsible for biomass accretion. First, the presence of emergent marsh vegetation is critical for maximizing accretion rates, so optimizing water depth for marsh colonization is vital; shallower wetlands (water depths of 25 cm) have been found to promote more complete colonization than deeper wetlands (55 cm). Second, rates of decomposition (which counteracts accretion) are minimized when the wetlands are flooded permanently (as opposed to temporarily or seasonally), since permanent saturation maintains anaerobic soil conditions. Finally, low rates of water flow are beneficial, since they minimize flushing and biomass loss from the wetlands.[47]

3 **Maximize the vertical accumulation of inorganic matter (where appropriate)**

Subsidence can also be reversed through the accumulation of inorganic (or "mineral") sediment. First, inorganic sediment has the potential to supplement the accumulation of organic matter as part of the marsh accretion process. Specifically, the rate of peat formation in the Delta is strongly related to periodic deliveries of sediments and accompanying nutrients, which increases inorganic sedimentation and organic matter accumulation.[48] Additionally, however, marshes in some parts of the Delta naturally contain peat with high bulk density and high proportions of inorganic content. These are generally high-energy areas proximal to the Delta's primary sediment sources (the major rivers and their distributaries in the upstream parts of the system) that are strongly influenced by fluvial and watershed processes. In contrast to the accumulation of organic matter, inorganic sedimentation increases with the depth of water over the marsh plain.[49] Accretion rates are also driven by suspended sediment concentrations, which varies spatially based on various regional and local conditions. Deposition increases when water is able to spread out and slow down (as it does on a floodplain). Landscape position should be a major factor in determining the relative importance of inorganic versus organic matter· and, thus, the best management regime for maximizing accretion rates.

4 **Once elevations are recovered, follow guidelines for re-establishing processes in intertidal areas**

The guidelines for re-establishing tidal marsh processes in areas at intertidal elevations (p. 46) also apply to re-establishing marsh processes in subsided areas, once intertidal elevations have been recovered. While not all are actionable until intertidal elevations are reached, these guidelines should be considered and anticipated during early planning phases of any reverse subsidence project:

- *"Marshes should experience full tidal action"* (p. 46, Guideline #1)
- *"Tidal marshes should have complex and variable patterns of tidal inundation"* (p. 46, Guideline #2)
- *"Tidal marshes should maintain processes that allow them to keep their extent over time"* (p. 46, Guideline #3)

LANDSCAPE CONFIGURATION & SCALE GUIDELINES

5 **Follow guidelines for re-establishing tidal marsh processes in intertidal areas**

The landscape configuration and scale guidelines for re-establishing tidal marsh processes in areas at intertidal elevations (pp. 46-47) also apply for re-establishing marsh processes in subsided areas, once intertidal elevations have been recovered:

- *"Tidal marshes should be as large as possible"* (p. 46, Guideline #4)
- *"Distance between tidal marshes should be minimized"* (p. 47, Guideline #5)
- *"The ratio of core to edge habitat should be maximized"* (p. 47, Guideline #6)
- *"The ratio of marsh to open water should increase"* (p. 47, Guideline #7)
- *"Maximize tidal marsh-water edge length through the development of interior channel networks"* (p. 47, Guideline #8)

6 **Minimize the distance from subsidence reversal projects to freshwater inputs with high sediment supplies**

Subsidence reversal might be most successful in areas of the Delta with high suspended sediment concentrations (proximal to major rivers and their distributaries). In these areas, inorganic sediment has the potential to supplement the accumulation of organic matter to promote marsh accretion. More research is needed to identify the response of freshwater marshes to sea-level rise and areas where marsh resilience might be improved via inorganic sediment deposition.

Sherman Island, 2014, photograph courtesy Gavin McNicol (UC Berkeley)

Managed wetlands (tule farm) on Sherman Island. Similar managed wetlands established on Twitchell Island have recovered lost elevation at an average rate of 4 cm/year.

Re-establish tidal zone processes in **channels and flooded islands**

SUPPORTED FUNCTIONS

marsh wildlife
(see pp. 88-91)

waterbirds
(see pp. 96-99)

fish
(see pp. 84-87)

biodiversity
(see pp. 106-107)

productivity
(see pp. 104-105)

The abiotic qualities of aquatic environments generally vary across time and space. These environmental variables interact at multiple scales to define the patch structure of the aquatic "landscape."[51] They influence or define the extent, quality, and availability of habitat, and have a strong effect on ecological processes. This strategy entails re-establishing hydrodynamics in tidal aquatic habitats that support desired ecological processes. In general, this means increasing the spatial and temporal heterogeneity of aquatic habitats, which have likely been homogenized over time.[52] It also includes management of novel ecosystems like flooded islands.

Since landscape morphology strongly affects the physical heterogeneity of aquatic habitats,[53] any attempt to establish desired tidal processes must carefully consider the Delta's geometry. Natural dendritic blind channels, for example—which were abundant in the historical Delta but have since decreased in length by 75–90%[54]—have a complex planform geometry that mediates interactions between land, water, and air, with important physical and ecological consequences.[55] A major component of this strategy is thus the creation of dendritic blind tidal channels that are embedded within and drain tidal marshes.

With continued sea-level rise, land subsidence, and levee failure events, the area and volume of flooded islands in the Delta is likely to increase in the future, with possible effects on salinity gradients and tidal range.[56] Flooded islands are currently favorable habitat for non-native freshwater lake fish, such as largemouth bass, that are valued for recreational fishing. Though remnant levees around flooded islands provide some shallow water habitat and wave barriers, the deep, open-water habitats that dominate the islands, are generally not well suited for native species, and are hazardous for migrating species.[57] However, the potential for these areas to improve conditions for native species if novel approaches are applied should not be ruled out.[58]

Specific tactics for implementing this strategy could include: recreating blind channel networks embedded in marshes; managing non-native species in flooded islands with flow barriers; enhancing aquatic migration corridors; and managing water pumps.

Image courtesy Google

tidal flows

First Mallard Branch
in Suisun Marsh

Dendritic channel, First Mallard Branch. Despite being the dominant historical channel form, there are no significant un-leveed dendritic blind channels embedded within marshes remaining in the Delta. This image shows First Mallard Branch in Suisun Marsh, one of the few remnant networks with a natural plan-form geometry in the upper estuary.

Ship Channel and surroundings: Though the Ship Channel and Toe Drain already function as blind channels with elevated residence times, both channels lack significant complexity, branching forms, and connection to marshes. The channels are both embedded within large areas of land situated at intertidal elevations, which creates possibility for the restoration of adjacent tidal marshes with low-order blind channels. Shoals within the Ship Channel benefit native fish.[59]

Northwest Delta: Cache Slough and Lindsey Slough present perhaps the best opportunity to improve tidal processes in open-water environments. Already these blind channels are longer than the tidal excursion (which promotes local turbidity maxima),[60] have some branches, and are lined with some tidal marsh. The channels are embedded within a large swath of leveed land that is at intertidal elevation—a major opportunity to re-establish tidal inundation regimes on low-order channel branches. Channel complexity and tidal hydrodynamics might further be improved through the removal or reconfiguration of Calhoun and Hastings cuts.

Flooded islands: are the result of subsided islands which have been permanently flooded, and are exposed to tidal action. These open-water "lakes" of the Delta, such as Frank's Tract and Mildred Island (3–5 m deep) tend to be less favorable to native species, and species such as non-native largemouth bass and sunfishes are abundant, as is invasive aquatic vegetation.[61] It is likely that continued subsidence and levee failures will lead to larger areas of deep, still-water habitat in the future, especially in the subsided islands of the Central Delta. Liberty Island, however, is a shallower flooded island (generally <1 m), at the downstream end of the Yolo Bypass, and at the edge of the historical tidal marsh. Liberty Island tends to be more turbid, due to higher wind-wave energy, and is largely devoid of submerged aquatic vegetation that supports non-native fishes.[62] It is possible that islands along the eastern and western edge of the Delta that are not very subsided, if flooded permanently, could function in the same way.

Landscape considerations
to re-establish tidal zone processes in channels and flooded islands

This map highlights the Delta's waterways—most of which are tidal—and shows their relation to both existing marshes and areas that are currently at intertidal elevation or will be in the relatively near future (with 6 ft [1.8 m] SLR). These may be areas where blind channels could be created, or tidal processes expanded.

Sacramento River

Sacramento

Eastern Delta: Numerous long but simplified blind channels extending eastward from the Central Delta intersect lands at intertidal elevation near their tips, where there is potential for restoring tidal inundation and some channel complexity. Cuts between individual meanders and whole networks likely reduce aquatic heterogeneity. Land is currently in agricultural use. There is little public/protected land and no restoration planned.

Suisun Bay

South Delta: Very few remnant blind tidal channels or tidally influenced distributaries extend off of the three primary branches of the San Joaquin River, though sizeable tracts of land situated at intertidal elevations present some physical opportunity for restoration. If land were acquired for this purpose, possible benefits for aquatic organisms could be tempered by entrainment risk and water-quality concerns (e.g., low dissolved oxygen, reverse flows). There is currently no public/protected land that could be considered for this strategy.

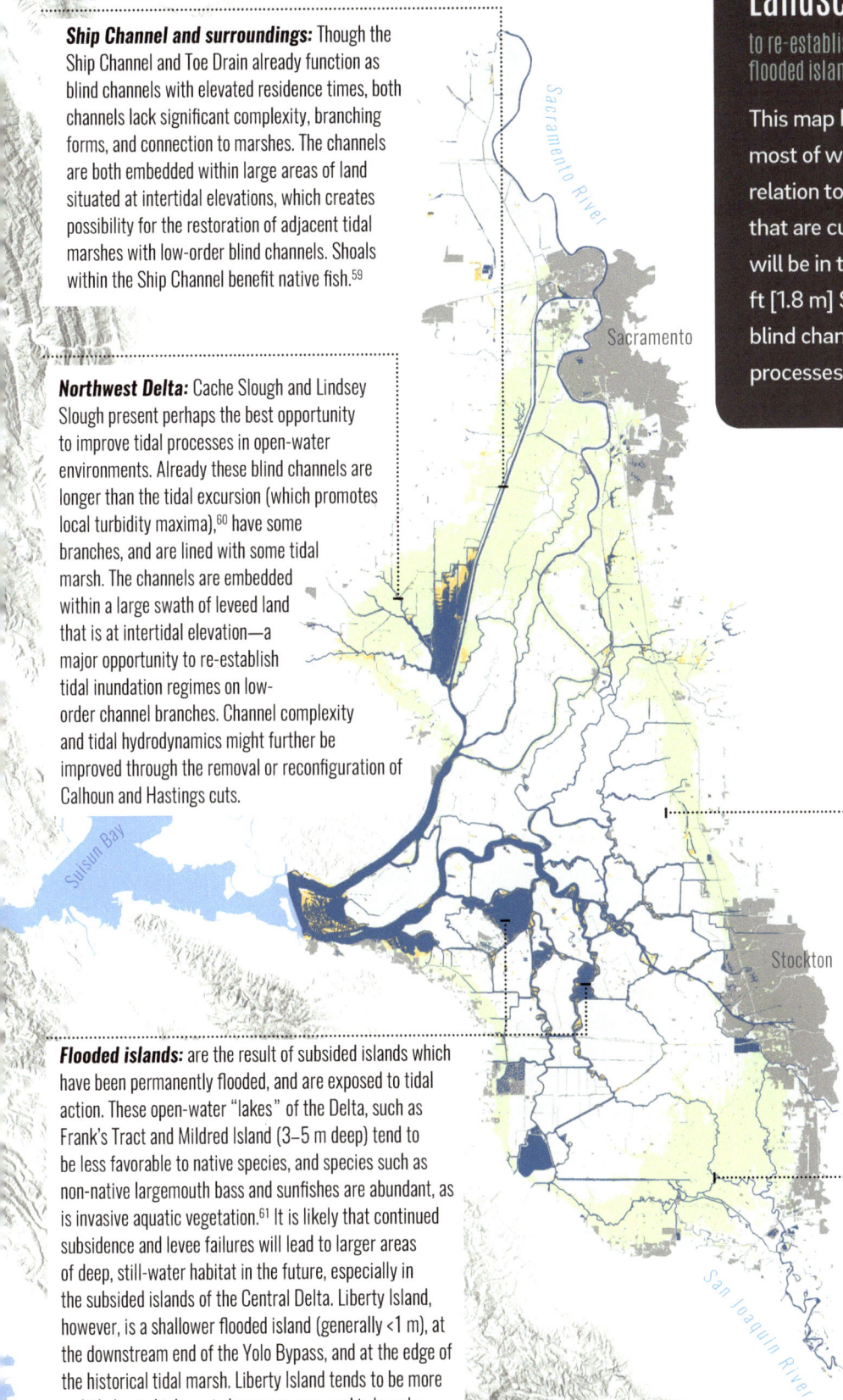

Stockton

San Joaquin River

N

10 miles

KEY DATA LAYERS

Intertidal elevation
current plus six additional feet

Existing marshes

Urbanized areas

SFEI AQUATIC SCIENCE CENTER

PHYSICAL PROCESS GUIDELINES

1 **The extent and duration of land-water connectivity should increase**

Connectivity between tidal channels and tidal marshes is critical for the exchange of energy, nutrients, organic and inorganic material, and organisms. Areas subject to tidal inundation have decreased by nearly 98%.[63] Though land and water were once connected through intricate dendritic channel networks, most channels are now separated from tidal marshes by flood-control levees. This change has likely reduced habitat heterogeneity since dendritic channel networks embedded in tidal marshes contribute to gradients in physical variables like residence time, velocity, and turbidity.[64] Via advection and concentration, smaller branches of dendritic channel networks may also provide resource subsidies from tidal marshes to organisms that are restricted to larger channels.[65] Tactics for accomplishing this strategy include the creation of new dendritic channels that are embedded within and drain restored tidal marshes. Tidal energy is constrained in the Delta, so further study is needed to understand the cumulative impact of future restoration projects on tidal range.

2 **Improve spatial and temporal variability of abiotic conditions**

With the loss of dendritic channel networks to reclamation, the widening and deepening of channels, and the creation of channel cuts, aquatic habitats in the Delta have become more homogenous.[66] The distance an individual fish, for example, must travel to experience different conditions has lengthened.[67] Changes in network topology that increase the connectivity of a system, such as channel cuts, can make it easier for disturbances to be transmitted through the network, resulting in more tightly correlated extinction risks for organisms in different parts of the system.[68] It is conceivable, for example, that increased hydrologic connectivity in the Delta has facilitated the spread of invasive aquatic organisms like the overbite clam and Brazilian waterweed. It may be possible to reduce the over-connectedness of aquatic habitats and to regain some level of habitat heterogeneity through the careful use of physical barriers. These could be positioned at the sites of channel cuts, effectively limiting the influence of artificial hydrologic connections that were created during the reclamation era. As described above, the restoration of dendritic tidal sloughs would also be expected to enhance habitat heterogeneity by generating gradients in abiotic conditions.

3 **Improve water quality in the tidal zone**

Water quality in the Delta is threatened by a mix of legacy sources, such as mercury mining, and new challenges, such as nutrients, pesticides, and contaminants of emerging concern.[69] Possible recommendations include reducing contaminant loads in agricultural and urban runoff, limiting high water temperatures in the summer using vegetative shading and connections to groundwater, and employing evaporative cooling, which takes places on marsh surfaces.

Lindsey Slough. The upper end of Lindsey Slough is one of the only remaining places in the Delta where a long blind tidal channel is embedded within a sizeable tidal marsh. This landscape configuration and the attendant tidal processes are critical for supporting tidal ecosystem functions.

Photograph courtesy Steve Culberson (USFWS)

4

Tidal channels should be embedded within tidal marshes

Lateral connectivity between channels and tidal marshes is critical. In addition to enhancing the aquatic food web, the ebb and flow of water onto the marsh plain through dendritic channels also affects heterogeneity in aquatic physical variables (e.g., temperature).[70]

> **88%** = **historical** percentage of blind channels (by length) embedded within marsh[71]
>
> **25%** = **modern** percentage of blind channels (by length) embedded within marsh[72]

5

The ratio of blind channel length to flow-through channel length should increase

Blind and flow-through channels differ in their environmental conditions and in the functions they provide for Delta wildlife. Blind, dendritic channels that terminate within wetlands serve as the capillary exchange system between wetland and aquatic areas, provide slow-moving water for energetic refugia, and create spatial complexity in habitat conditions.[73] A 75–90% decrease in blind channel length and a slight increase in the length of flow-through channels (from channel cuts) has yielded a dramatic decrease in the ratio of the two channel types.[74] Increasing the blind to flow-through channel length ratio should help to increase hydrodynamic complexity and to recover lost processes and functions.

> **4.25** = **historical** ratio of blind to flow-through channel length[75]
>
> **0.44** = **modern** ratio of blind to flow-through channel length[76]

6

Blind tidal channels should have an appropriate branching geometry

Blind tidal channels naturally develop a dendritic, branching form, with multiple channel orders (generally up to four in the historical Delta), which should be restored to benefit aquatic and marsh wildlife. This geometry increases the length of edge between marsh and channel habitats; contributes to heterogeneity in physical variables, such as water residence time, depth, temperature, turbidity, and velocity;[77] affects predator-prey dynamics by influencing predator search patterns; sets up resource subsidies from small/shallow branches to larger/deeper ones; and can enhance the size, diversity, and resilience of resident wildlife where branches intersect at nodes.[78]

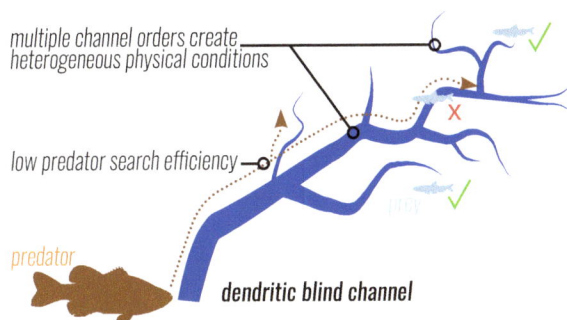

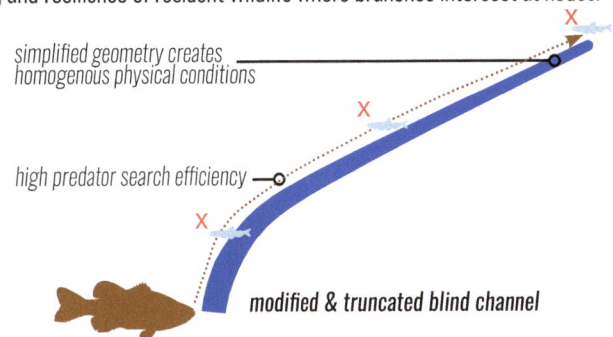

multiple channel orders create heterogeneous physical conditions

low predator search efficiency

predator

dendritic blind channel

simplified geometry creates homogenous physical conditions

high predator search efficiency

modified & truncated blind channel

7

Blind channels should be longer than the tidal excursion length

Many blind channels in the Delta today (including ones constructed as part of restoration projects) are significantly shorter than historical blind channels and local tidal excursion lengths. Blind channels that are longer than the average tidal excursion length are needed to generate residence-time gradients along the length of the channel.[79] Long, multi-order tidal channels will require relatively large tracts of marsh in which to become established (see p. 46).

Re-establish connections between **streams and tidal floodplains**

SUPPORTED FUNCTIONS

marsh wildlife
(see pp. 88-91)

edge wildlife
(see pp. 100-103)

waterbirds
(see pp. 96-99)

riparian wildlife
(see pp. 92-95)

fish
(see pp. 84-87)

biodiversity
(see pp. 106-107)

productivity
(see pp. 104-105)

The fluvial-tidal zone is found at the upstream-most edge of regular tidal inundation and the transition from tidal to non-tidal environments.[80] This zone is dynamic, shifting with seasonal and inter-annual variation in tidal range and freshwater input.[81] The fluvial-tidal zone exhibits a gradient in hydraulic conditions as river flows meet incoming tides, resulting in changing channel geomorphic characteristics[82] and often a change in species composition and behavior.[83] Sediment is transported upstream by flood tides and downstream by ebb tides and river flows, such that a reach of high turbidity and sediment deposition can occur where tidal and fluvial flows meet and slow. The tidal-fluvial transition zone is also the region where upstream woody riparian habitats naturally transition into non-tidal and tidal freshwater emergent wetlands downstream, creating a range of habitats and important ecotones outside the channel itself. The potential importance of re-establishing habitats in this zone is illustrated by research into the downstream migrations of juvenile Chinook, which has suggested that if the tidal-fluvial transition zone occurs where habitat conditions are relatively good, including where predator densities are relatively low, the fish are likely to experience lower predation mortality and perhaps improved rates of growth.[84]

To restore the fluvial-tidal transition zone, we must re-establish connections between streams and tidal floodplains, thereby recreating deltas within the Delta. Specific tactics for implementing this strategy include the creation of distributaries at the upstream end of tidal islands through levee breaches; the creation of marshes along rivers at the upper reaches of tidal influence through levee breaches; managing flows to situate the tidal-fluvial transition zone in areas with desirable habitat conditions;[85] and the restoration of riparian processes and off-channel habitats along channels in this zone (especially on remnant natural levee topography).

Image courtesy Google

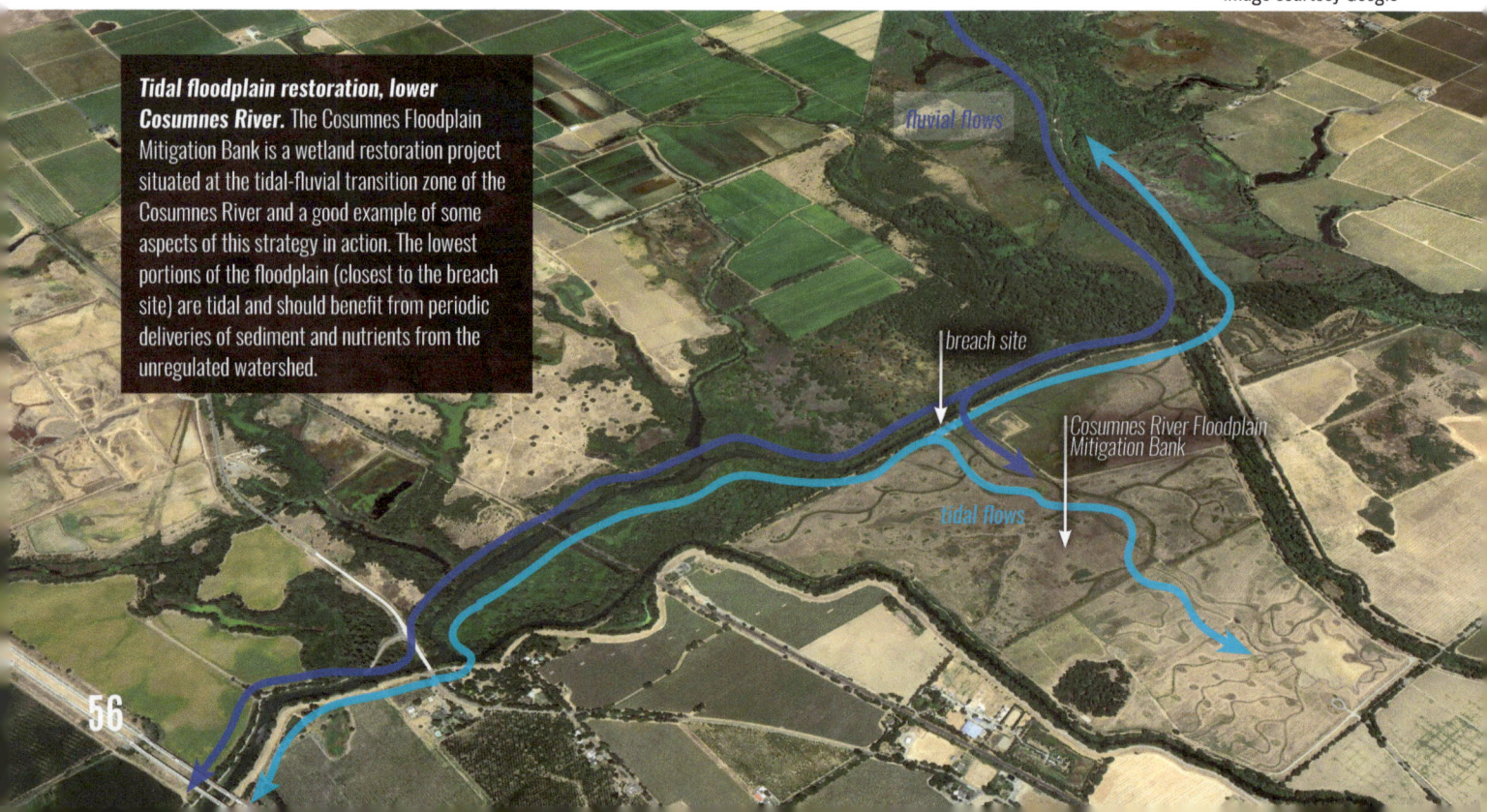

Tidal floodplain restoration, lower Cosumnes River. The Cosumnes Floodplain Mitigation Bank is a wetland restoration project situated at the tidal-fluvial transition zone of the Cosumnes River and a good example of some aspects of this strategy in action. The lowest portions of the floodplain (closest to the breach site) are tidal and should benefit from periodic deliveries of sediment and nutrients from the unregulated watershed.

fluvial flows

breach site

Cosumnes River Floodplain Mitigation Bank

tidal flows

Landscape considerations
to re-establish connections between streams and tidal floodplains

This map highlights 1) the Delta's major fluvial inputs and 2) areas that are currently at intertidal elevation or will be in the relatively near future (with 6 ft [1.8 m] SLR). Areas where incoming rivers and streams intersect this intertidal elevation band should be considered opportunity areas for re-establishing connections between streams and tidal floodplains.

Putah Creek: Flow and sediment from moderate-size watershed regulated by Berryessa Dam. Extensive land at creek's fluvial-tidal interface within the Yolo Bypass. Potential location for recreating historical distributary network ("sinks") at distal ends of west side tributaries (instead of connecting directly to the Toe Drain).

Yolo Bypass / Sacramento River: Yolo Bypass receives flood waters from Sacramento River. Flows regulated by upstream dams and Fremont Weir. Sediment supply relatively high but also impacted by dams. Moderate amounts of protected land. Major restoration planning efforts underway.

Morrison Creek: Small watershed, but unregulated by dams. Some existing protected land in fluvial-tidal transition zone. Note that this is the northerly limit of the tides east of the Sacramento River, so tidal energy may be limited..

Cosumnes River / Mokelumne River: The Cosumnes River is the largest unregulated tributary of the Delta. The Mokelumne River has a relatively large watershed but is regulated. There is extensive protected land in the fluvial-tidal transition zones of these streams, with good examples of this strategy already in action; restoration efforts could still be expanded.

Marsh Creek: Relatively small watershed is regulated by Marsh Creek Dam. Extensive protected areas (Big Break and Dutch Slough) at fluvial-tidal interface. May have issues with mercury contamination.

Bear Creek: Small watershed, but unregulated by dams. No existing protected land in fluvial-tidal transition zone.

San Joaquin River: Large, but regulated, watershed. Sediment supply lower than Sacramento River but still potentially substantial. Almost no protected land within fluvial-tidal transition zone.

Dutch Creek: Small watershed, but unregulated by dams. No existing protected land in fluvial-tidal transition zone.

KEY DATA LAYERS

Intertidal elevation[10]
- currently intertidal
- currently intertidal +3 ft (0.9 m)
- currently intertidal +6 ft (1.8 m)

Waterways

N

10 miles

SFEI AQUATIC SCIENCE CENTER

PHYSICAL PROCESS GUIDELINES

1 **Restore flooding regimes with a magnitude, timing, extent, and duration of inundation that supports desired ecological functions**

Aquatic habitats in the historical Delta were complex and dynamic, providing many resources and opportunities for native species. Historically, the Delta exhibited wide seasonal and interannual variation in flood patterns. There was variety in tidal and fluvial inundation, including its depth, duration, timing, and extent. The areas where flows transitioned from being tidally dominated to fluvially dominated were particularly dynamic, supporting complex variation in food resources, abiotic conditions, and channel/off-channel habitats (including tidal and non-tidal freshwater marshes). Tactics for increasing the complexity and variability of flood regimes to support native species in the tidal-fluvial transition zone overlap with tactics to restore fluvial zone processes, and include managing reservoir releases to provide functional flows and flooding at particular times of the year.[86] Shifts in the rain/snow ratios of the upper watershed precipitation, especially in the San Joaquin basins, may increase flood flows and change the timing of flooding.[87] Climate-driven flooding and managed floods will also require the creation on levee setbacks to accommodate those flows and allow floodplains to grade into marsh plains, thus reconnecting the gradient between fluvial and tidal processes. It is equally important to increase opportunities for seasonal, short-duration flooding and regular tidal inundation.[88] Overall, processes within the tidal-fluvial transition zone should be restored and also managed to maintain complex flooding patterns and resultant habitat mosaics.

2 **Increase delivery of flows with high suspended sediment concentrations to marshes in the fluvial-tidal transition zone**

Tidal-fluvial transition zones are areas of high sediment deposition, and may be areas where restored marshes can achieve necessary accretion rates to keep pace with sea-level rise. Allowing tides to fully inundate restored tidal marshes will bring suspended sediment to newly restored marshes. Weirs and levee notches often "decant" flows, allowing only the sediment-poor top-layer of water to flow from rivers onto adjacent marshes. Allowing sediment-rich flows to inundate fluvial floodplains, including marsh restoration projects, and increasing residence time of floods on these areas will allow for more settling to occur and increase sediment deposition rates. When flood flows are quickly routed off marshes and floodplains, less deposition occurs.

3 **Prepare for migration of the fluvial-tidal zone upstream with sea-level rise**

When planning the fluvial-tidal zone of the future, managers must consider the likely landward migration and compression of this zone due to sea-level rise.[89] Increased flashiness and changed timing and intensity of flows will also contribute to the uncertainty around the location and extent of this zone.[90] Guidelines include: acquiring parcels on either side of a channel upstream of the current fluvial-tidal interface to allow for accommodation space for increased inundation and upstream migration of marshes; setting back levees upstream of the fluvial-tidal zone in anticipation of this increased inundation; and accounting for biodiversity within this transition zone that may be lost or need to move with climate change.[91]

LANDSCAPE CONFIGURATION & SCALE GUIDELINES

4 **Increase the extent of productive ecotones by restoring woody riparian habitat adjacent to marshes**

Historically, along the vast majority of their length, the Delta's elevated woody riparian corridors graded down to marshes. Continuous edges between marshes and woody riparian habitats provide riparian wildlife with access to wetland habitats for foraging, and provide wetland wildlife with cover, high-water refuge, and alternative food sources. Marshes also dissipate flood waters that move through riparian habitats, reducing flood heights within the riparian corridor and the associated mortality of terrestrial animals like riparian woodrat and brush rabbit.[92] The length of the adjacency between marsh and valley foothill riparian habitats has decreased more than the adjacency between marsh and any other terrestrial habitat type.[93] Landscape-scale restoration should seek to recover some of the associated lost functions by restoring woody riparian habitats adjacent to tidal marshes at the fluvial-tidal interface.

> **89%** = **historical** woody riparian edge adjacent to marsh (excluding edge adjacent to water)[94]
>
> **15%** = **modern** woody riparian edge adjacent to marsh (excluding edge adjacent to water)[95]

6 **Follow certain guidelines for re-establishing marsh processes in intertidal areas**

Many of the guidelines for re-establishing tidal marsh processes in areas at intertidal elevations (pp. 46-47) also apply for re-establishing connections between streams and tidal floodplains (though in this zone the marshes in question could be tidal or non-tidal):

- *"Tidal marshes should be as large as possible"* (p. 46, Guideline #4)
- *"The ratio of core to edge habitat should be maximized"* (p. 47, Guideline #6)
- *"The ratio of marsh to open water should increase"* (p. 47, Guideline #7)
- *"Maximize tidal marsh-water edge length through the development of interior channel networks"* (p. 47, Guideline #8). Note that channel configurations in the tidal-fluvial transition zone will be different than they are in the tidal zone (likely less sinuous and more dynamic over time).

7 **Follow certain guidelines for re-establishing tidal zone processes in subsided areas**

Some of the guidelines for re-establishing tidal marsh processes in subsided areas (p. 51) also apply for re-establishing connections between streams and tidal floodplains. Specifically, efforts to restore the tidal-fluvial transition zone should *"Minimize the distance from subsidence reversal projects to freshwater inputs with high sediment supplies"* (p. 51, Guideline #6). These strategies have a positive synergy because many of the Delta's least subsided tracts with the greatest potential to recover intertidal elevation over time are found in the fluvial-tidal transition zone.

8 **Follow certain guidelines for re-establishing fluvial processes along streams**

The guidelines for re-establishing fluvial processes along streams (pp. 63-65) also apply for re-establishing connections between streams and tidal floodplains, insofar as both strategies entail re-establishing woody riparian corridors:

- *"Woody riparian habitats should be connected to the processes that maintain them"* (p. 63, Guideline #6)
- *"Woody riparian corridors should be as wide as is feasible"* (p. 63, Guideline #7)
- *"Woody riparian patches should be as large as is feasible"* (p. 64, Guideline #8)
- *"Minimize the length and frequency of gaps in woody riparian corridors"* (p. 65, Guideline #9)

Re-establish fluvial processes along **streams**

SUPPORTED FUNCTIONS

marsh wildlife
(see pp. 88-91)

edge wildlife
(see pp. 100-103)

waterbirds
(see pp. 96-99)

riparian wildlife
(see pp. 92-95)

fish
(see pp. 84-87)

biodiversity
(see pp. 106-107)

productivity
(see pp. 104-105)

Runoff from more than 40% of California's land area drains through the Delta, and out to San Francisco Bay and the Pacific Ocean.[96] The two main river systems—the Sacramento and San Joaquin—along with other major tributaries on the east and west sides of the Delta historically delivered freshwater, coarse and fine sediment, nutrients, and other materials to the Delta. These inputs supported a diverse array of habitat types, including non-tidal marshes in the riverine floodplains and riparian forests on elevated natural levees. The seasonal and interannual variability of the inputs created a portfolio of support for wildlife and other ecological functions.

As dams were built during the 20th century on the major rivers flowing from the Sierra Nevada mountains, coarse sediment and water that once flowed to the Delta was increasingly trapped or regulated in reservoirs. Downstream of the reservoirs and dams, rivers were starved of sediment, causing channel incision and subsequent decreases in groundwater levels, with negative impacts on riparian vegetation. Dams prevent anadromous fish access to the upper watershed and at least 72% of their former spawning and holding habitat.[97] The lack of large regular floods, combined with the building of artificial levees, disconnected rivers from their floodplains and decreased in-channel complexity.

A combination of dams and water diversions has caused Delta inflows to change dramatically, altering the influence of freshwater supply on the geometry and extent of the channels, habitats, and water quality of the Delta. Historically, an average of about 39.1 km^3 (31.7 million acre-ft) of runoff is estimated to have flowed into the Central Valley per year.[98] The flows were slowed by the adjacent floodplains, extensive non-tidal marshes, and dendritic channel network, and thus had higher residence times than the flows today that are shunted through the rip-rapped, flood-protection channels.[99] A mix of dams, diversions, and water exports have reduced historical Delta outflows by an average of 50% in recent years.[100]

Climate change is altering the timing and intensity of snowmelt and rainfall. Future runoff from the Sierra Nevada is projected to peak earlier in the year, owing to less precipitation falling as snow and more falling as rain that immediately runs off, along with an earlier melt of the snow that does accumulate.[101] This could affect both magnitudes and frequencies of floods by increasing the drainage areas from which rain runs off most rapidly and by increasing the intensity of large storms.[102] These changes will also result in lower summertime flows, increasing stresses on ecosystems, and potentially increasing the risk of fire.[103] Increasing human population will drive greater demand for summer runoff, possibly leading to more upstream storage and diversion. However, regulation of summer flows could counteract future changes.

Specific tactics for implementing this strategy include levee setbacks, reservoir releases that mimic naturalistic hydrographs, sediment augmentation/trucking, dam notching/slurries, bank set-backs/channel-to-floodplain reconnection, and targeted groundwater recharge to support non-tidal marshes. The Delta also has substantial remnant fluvial topographic features, such as natural levees along the rivers and depressions at the former sites of flood basin lakes that should be incorporated into plans for recovering fluvial processes and habitats.

Landscape considerations
to re-establish fluvial processes along streams

This map shows the Delta's 1) elevation, 2) major fluvial inputs and waterways, 3) floodplain habitats, and 4) riparian habitats. The elevation layer reveals remnant topographic features along streams, such as natural levees and flood basins that provide physical opportunities for re-establishing natural processes and habitats. Overlaying existing woody riparian habitats shows where there are gaps in the riparian corridor and potential for increasing connectivity.

Yolo Bypass: Opportunities to expand the Bypass, increase flood frequency, and improve fish passage are well-studied. Increasing the capacity of the Bypass could allow for an increase in riparian vegetation (especially at mouths of west-side tributaries) and in areas subject to long-duration inundation. Extensive protected land with multiple planning processes underway.

Sacramento River and distributaries: Despite extensive remnant natural levee topography, there are constraints to the establishment of functional woody riparian habitats along the Sacramento River, especially adjacent to urban areas. Some opportunities in parks along river. Also extensive remnant natural topography along distributary sloughs (e.g., Babel, Elk, Miner).

Northeast Delta: Remnant topography of a Sacramento River splay provides opportunity to expand existing narrow bands of woody riparian habitat (e.g., along Borrow Channel, Snodgrass Slough), with potential connectivity to downstream areas with wide woody riparian habitats and functional floodplains (e.g., the Cosumnes Preserve). Remnant floodplain lakes and other topographic lows present opportunities for re-establishing long-duration inundation and non-tidal wetlands associated with the Sacramento Basin. Extensive protected land.

Cosumnes River / Mokelumne River: Remnant natural levees along Mokelumne River, which intersects protected land in some places (McCormack-Williamson Tract, Cosumnes Preserve, Cosumnes Mitigation Bank). The Cosumnes River has a relatively natural flow regime, though the Mokelumne is regulated by multiple dams.

San Joaquin River: Extensive wide woody riparian habitats, but many notable gaps, especially in lower reaches of fluvial zone. Much remnant floodplain topography (natural levees, channel scrolls, depressions), but a pronounced lack of floodplain connectivity and associated habitats. Planned urban developments constrain some opportunity for woody riparian habitat and floodplain restoration. Very little existing protected area (and those areas that exist are agricultural conservation easements). Though the river is highly regulated, peak flows are predicted to increase with climate change.

Sacramento River

Suisun Bay

San Joaquin River

KEY DATA LAYERS

Elevation (NAVD88)
>7 m
<-1 m

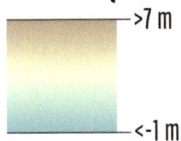

Existing woody riparian vegetation
- narrow
- wider than 100 m
- wider than 500 m

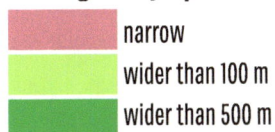

Fluvial floodplains
- short duration seasonal inundation

N

1:600,000

10 miles

SFEI AQUATIC SCIENCE CENTER

PHYSICAL PROCESS GUIDELINES

1 **Re-establish functional flows**

Establish reconciled "functional flows," focusing on components of the hydrograph that provide particular geomorphic or ecological functions.[104] Such a hydrograph includes wet-season initiation flows, peak magnitude flows, spring recession flows, dry season low flows, and interannual variability that reflect natural/historical conditions. This includes flooding that has the appropriate frequency, magnitude, duration, timing, and rate of change, or otherwise emulates the natural patterns of flow variability. A key tool could be using pulsing flows from reservoirs,[105] paired with landscape solutions such as bypass channels and floodplain widening.[106] Work has been done to develop environmental flow standards that quantify the amount of water needed to maintain the ecological integrity of rivers.[107] Criteria adopted by the State Water Resources Control Board state that, in order to protect public trust resources in the Delta and preserve the attributes of a naturally variable system to which native fish species are adapted, winter and spring flows through the Delta and into the Bay should be 75% of unimpaired flows.[108] In recent years, spring Bay inflows have averaged only 44% of unimpaired flows.[109]

2 **Establish appropriate flows and sediment transport/delivery to maintain in-channel habitats**

In order to maintain in-channel habitats, including macro-habitat features such as bars, pools, riffles, undercut banks, and wood jams, channels must receive sufficient flows for basic longitudinal connectivity along perennial streams, allowing for adult fish passage upstream,[110] and also likely, support for genetic diversity and population resilience. Flows that drive channel maintenance and habitat creation must be sufficient to transport coarse and fine sediment, sort sediment, and scour the bed locally, as well as drive bank erosion, channel migration, and, where appropriate, the creation of backwaters.[111] These flows are also critical to maintaining water-quality objectives, including salinity gradients and temperature,[112] as well as coherent chemical gradients for natal stream homing.[113]

3 **Establish appropriate flows and sediment transport/delivery to create, maintain, and activate complex off-channel, floodplain, and woody riparian habitats**

Flows should be sufficient to drive critical geomorphic floodplain processes, such as frequent overbank flooding and delivery of water, nutrients, and fine sediments; localized scour; and sediment deposition. These flows should have a long enough duration on the floodplain to support ecological functions and activate a food web. For example, inundation with a residence time of at least 2 days allows for phytoplankton production, while at least 14 days allows for zooplankton production and Chinook salmon rearing; even longer is required for Sacramento splittail spawning and rearing.[114]

4 **Increase the frequency of long-duration inundation events**

While historically the Yolo and Sacramento basins and the San Joaquin River floodplain would experience inundation persisting up to 6 months in wet years,[115] engineered levees and drainage systems have effectively eliminated long-duration fluvial flooding from the Delta. This has been shown to have negative consequences for species like the Sacramento splittail whose life-history strategy relies on long periods of inundation.[116] Studies have found that splittail need a minimum of 45 days of access to inundated floodplains for spawning, egg incubation, larval rearing, and migration downstream to the Delta.[117] During the past 75 years, the Yolo Bypass has been inundated for 45 days in only 10% of years. Under unimpaired conditions, it is thought that the Yolo Bypass would have flooded (with at least 10,000 cfs) for at least one month in 54% of years, and for at least two months in 26% of years.[118] Note, however, that some functions, like the production and export of productivity downstream, benefit from periodic connections and disconnections of floodplains with channels,[119] so some flood pulses within longer general periods of inundation might be beneficial.

5 **Allow for groundwater recharge to support hyporheic exchange and cold-water refuge**

The interaction between groundwater and surface water (hyporheic exchange) can provide "hotspots" of primary productivity resulting from the upwelling of nutrient-rich water to the surface, and can provide dissolved oxygen and organic matter to benthic invertebrates.[120] Hyporheic exchange can also cool and stabilize local surface-water temperatures, creating refuge for fish and other sensitive organisms.[121] A high groundwater table also supports non-tidal freshwater emergent wetlands in riverine flood basins.[122] Although groundwater is generally high in the Delta, there are notable cones of depression at the Delta's perimeter due to localized groundwater pumping. These areas often coincide with the landward tips of blind channels and other streams that may have once been, but no longer are, supplied with groundwater.

LANDSCAPE CONFIGURATION & SCALE GUIDELINES

6 **Woody riparian habitats should be connected to the fluvial processes that maintain them**

Periodic deliveries of water and sediment are required to maintain the environmental conditions (e.g., moisture gradients and groundwater levels) and geomorphic surfaces (e.g., natural levees and point bars) that sustain woody riparian habitats and their associated functions.[123] However, due to the construction of artificial levees and associated habitat loss, much of the woody vegetation found in the Delta today is not actually connected to beneficial flooding and the larger set of fluvial processes that form and maintain woody riparian habitats over time. A simple first step for re-establishing such functional woody riparian habitats (subject to the processes that maintain them) is to ensure that woody riparian patches are adjacent to streams.

7 **Woody riparian corridors should be as wide as is feasible**

Woody riparian corridors should be as wide as is feasible, since the functions these habitats support increase with width. Though relatively narrow riparian corridors can provide some ecological functions (e.g., corridors at least 5–25 m wide are needed to ensure leaf litter input to streams), the full suite of ecological functions provided by riparian habitats is only achieved at much greater widths (e.g., optimal nesting habitat for Western Yellow-billed Cuckoo is at least 600 m wide). Wide woody riparian corridors have been disproportionately lost over time.

25 m	= minimum corridor width for leaf litter input to streams[124]
80 m	= minimum corridor width for large woody debris input to streams[125]
200 m	= minimum corridor width "suitable" for Western Yellow-billed Cuckoo nesting[126]
>600 m	= corridor width "optimal" for Western Yellow-billed Cuckoo nesting[127]
53%	= percentage of **historical** woody riparian habitats (by length) wider than 200 m[128]
5–8%	= percentage of **modern** woody riparian habitats (by length) wider than 200 m[129]
20%	= percentage of **historical** woody riparian habitats (by length) wider than 600 m[130]
<1%	= percentage of **modern** woody riparian habitats (by length) wider than 600 m[131]

8 Woody riparian patches should be as large as is feasible

Woody riparian habitat patches should be as large as physically feasible, since the functions supported by woody riparian areas increase with patch size. Riparian-corridor width and patch size are dependent in unconstrained systems on physical variables, like the size of the stream and its sediment load.

4 ha	= minimum patch size for Yellow-breasted Chat[132]
20–80 ha	= "marginal" for Western Yellow-billed Cuckoo nesting[133]
>80 ha	= "optimal" for Western Yellow-billed Cuckoo nesting[134]
862 ha	= mean **historical** woody riparian patch size (SD = 2,785)[135]
6 ha	= mean **modern** woody riparian patch size (SD = 45)[136]

Image courtesy Google

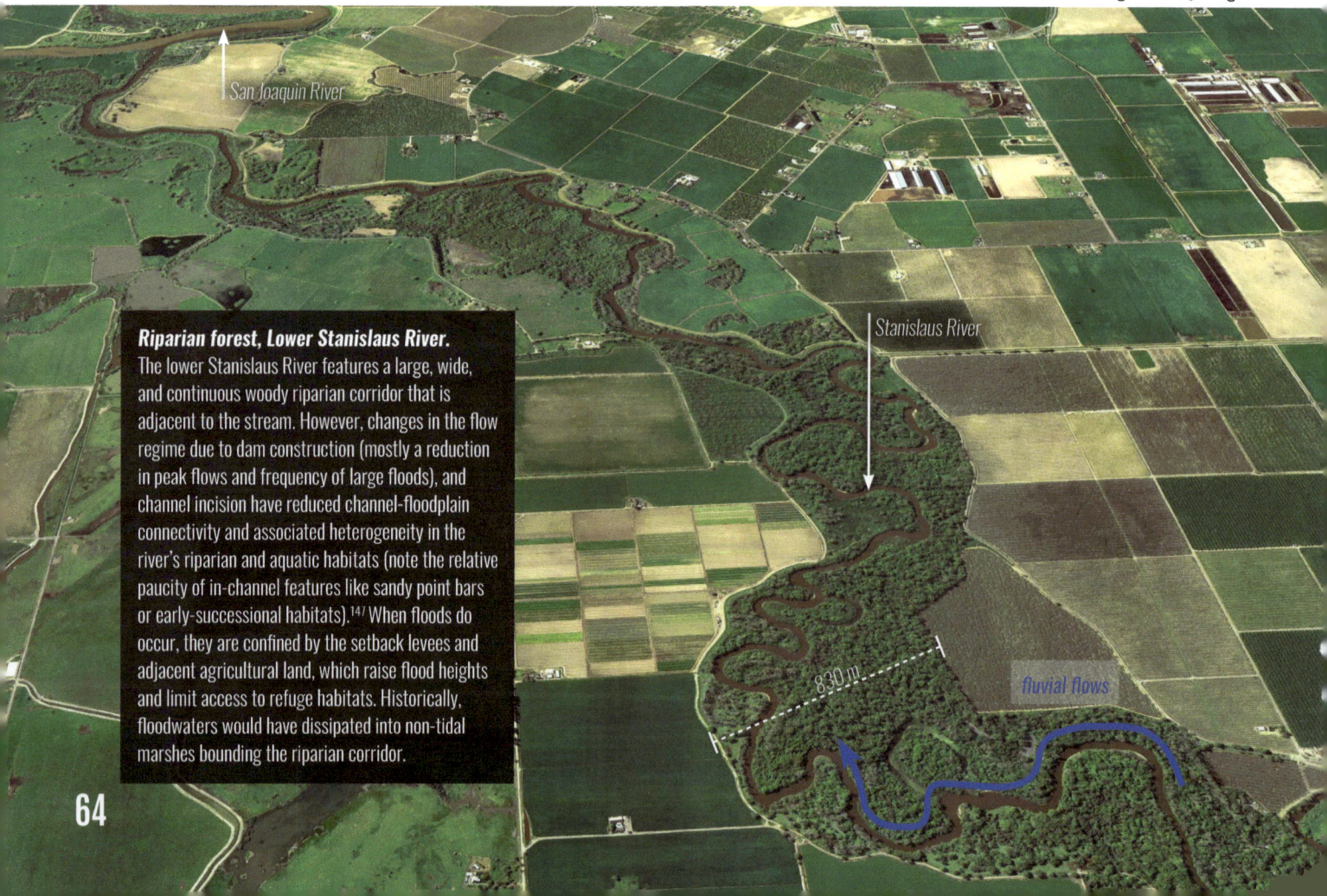

Riparian forest, Lower Stanislaus River.
The lower Stanislaus River features a large, wide, and continuous woody riparian corridor that is adjacent to the stream. However, changes in the flow regime due to dam construction (mostly a reduction in peak flows and frequency of large floods), and channel incision have reduced channel-floodplain connectivity and associated heterogeneity in the river's riparian and aquatic habitats (note the relative paucity of in-channel features like sandy point bars or early-successional habitats).[147] When floods do occur, they are confined by the setback levees and adjacent agricultural land, which raise flood heights and limit access to refuge habitats. Historically, floodwaters would have dissipated into non-tidal marshes bounding the riparian corridor.

San Joaquin River

Stanislaus River

830 m

fluvial flows

9   Minimize the length and frequency of gaps in woody riparian corridors

While the historical Delta was characterized by long, continuous corridors of woody riparian habitats, the modern Delta is characterized by fragmented woody riparian corridors with sizeable and frequent gaps. This is problematic, since gaps in corridors (even small ones) can decrease landscape permeability for riparian wildlife, creating barriers to their movement.[137] These barriers, in turn, can reduce the resilience and long-term persistence of populations.[138] Reducing the size and frequency of gaps should improve conditions for riparian wildlife. The reference values below reflect some of the best available science from other regions that may be analogous for Delta species.

13 m = woody riparian gap size known to decrease landscape permeability for salamanders (Appalachian Mountains)[139]
45 m = gap size known to decrease likelihood of forest songbird movement by half (Canada)[140]
100 m = typical maximum gap crossing distance of dispersing songbirds (North America)[141]
Refer to maps in the companion *Detla Transformed* report to see gap sizes in historical and modern Delta[142]
55 km = longest **historical** continuous stretch of woody riparian habitat (Delta)[143]
16 km = longest **modern** continuous stretch of woody riparian habitat (Delta)[144]

10   Follow guideline for re-establishing connections between streams and tidal floodplains

One of the landscape configuration and scale guidelines for re-establishing connections between streams and tidal floodplains also applies to re-establishing fluvial processes along streams:

- *"Increase the extent of productive ecotones by restoring woody riparian habitat adjacent to marshes"* (p. 59, Guideline #4)

11   Floodplains should generally have connections at both upstream and downstream ends

Floodplains and flood basins historically received floodwaters, either through bank overtopping, or splays in natural levees.[145] Floodwaters then receded off the floodplain and drained over the course of the dry season, a process that extended channel-floodplain connectivity and increased the exchange of water, sediment, and nutrients. For migratory fish, connections to the channel network at both the upstream and downstream end allowed floodplains to serve as (highly productive) migratory pathways.[146] In the modern Delta, with increased concerns about the stranding of endangered species, it is important to design floodplains and bypasses to drain back into the river system through channels, allowing for fish passage at both ends.

Re-establish **tidal-terrestrial transition zone processes**

SUPPORTED FUNCTIONS

marsh wildlife
(see pp. 88-91)

edge wildlife
(see pp. 100-103)

waterbirds
(see pp. 96-99)

riparian wildlife
(see pp. 92-95)

biodiversity
(see pp. 106-107)

productivity
(see pp. 104-105)

The tidal-terrestrial transition zone, or "T-zone," is the area of interactions between tidal and terrestrial processes that result in mosaics of habitat types, assemblages of plant and animal species, and sets of ecosystem services that are distinct from those of the adjoining estuarine or terrestrial ecosystems.[148] Though not all tidal marshes in the historical Delta bordered terrestrial habitats (some Central Delta marshes were completely surrounded by aquatic areas), the Delta did feature more than 500 km (320 miles) of tidal-terrestrial edge.[149] This lengthy T-zone extended around the tidal Delta in a near-continuous ring, and also made sizeable inroads towards the Delta's interior along elevated natural levees. At least nine distinct terrestrial habitat types (valley foothill riparian, willow riparian scrub or shrub, willow thicket, wet meadow and seasonal wetland, alkali seasonal wetland complex, vernal pool complex, grassland, oak woodland/savanna, and stabilized interior dune vegetation) were found at the terrestrial edge of the Delta's tidal marshes, meaning the T-zone and the functions it provided were quite diverse.[150] Some T-zone types (e.g., marsh to wet meadow) would generally have been broad and gradual, while others would have been narrower and more clearly delineated (e.g., marsh-stablized interior dune).

The extent and distribution of the T-zone is highly sensitive to land-use change since it can be eliminated by the loss of either tidal marsh or the adjacent terrestrial habitat type—in other words, both "ends" of the gradient are required for a functional T-zone. This strategy can therefore be thought of as the coordinated implementation of both the strategies to restore tidal zone processes (pp. 44-55) and the strategy to re-establish connected terrestrial habitats (pp. 70-73). Fundamental to this strategy is the recovery of key processes, namely tidal inundation over lands at intertidal elevation and the simultaneous re-establishment of terrestrial hydrology (both surface and subsurface flows). Since T-zones will shift upslope as sea levels continue to rise, their re-establishment should be prioritized in areas with sufficient accommodation space.

Image courtesy Google

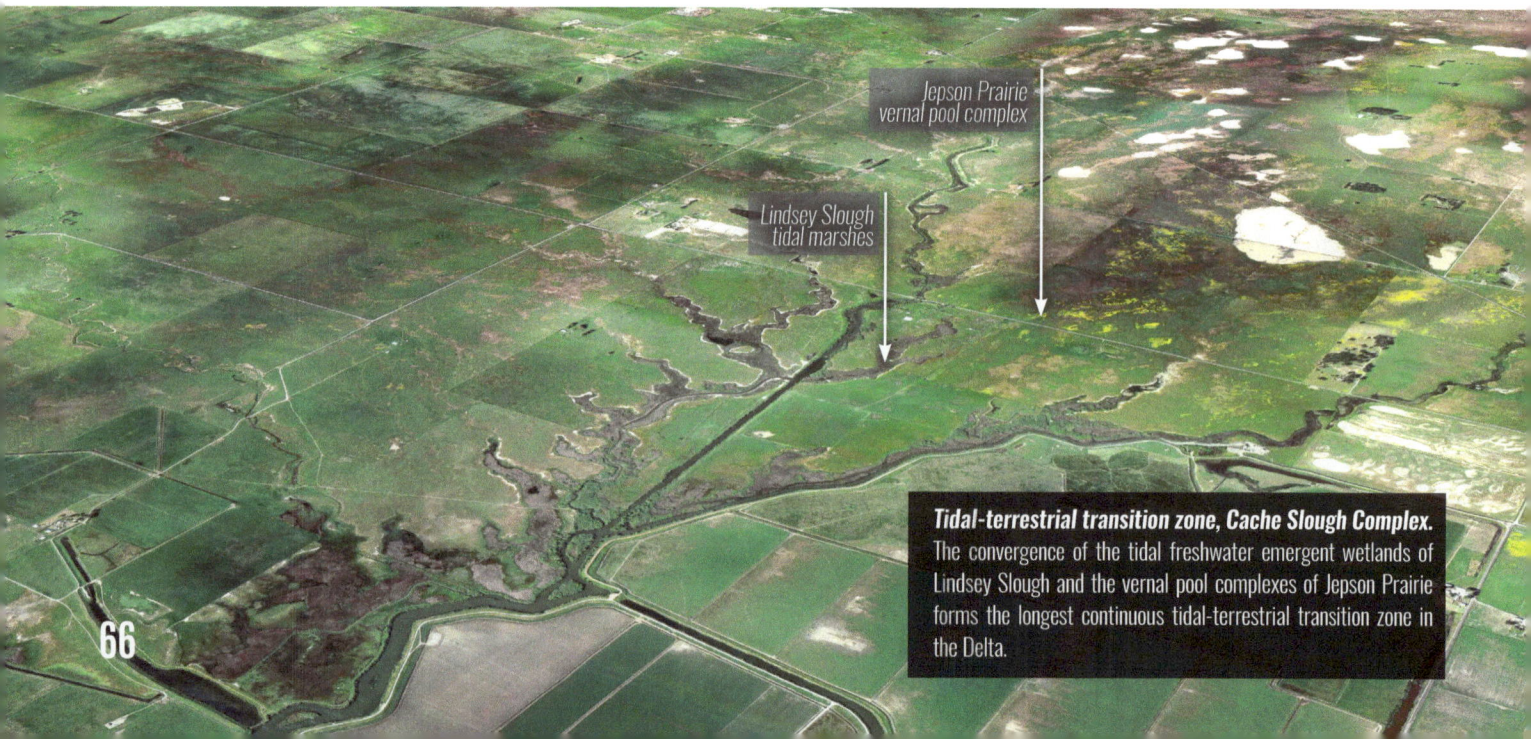

Jepson Prairie
vernal pool complex

Lindsey Slough
tidal marshes

Tidal-terrestrial transition zone, Cache Slough Complex.
The convergence of the tidal freshwater emergent wetlands of Lindsey Slough and the vernal pool complexes of Jepson Prairie forms the longest continuous tidal-terrestrial transition zone in the Delta.

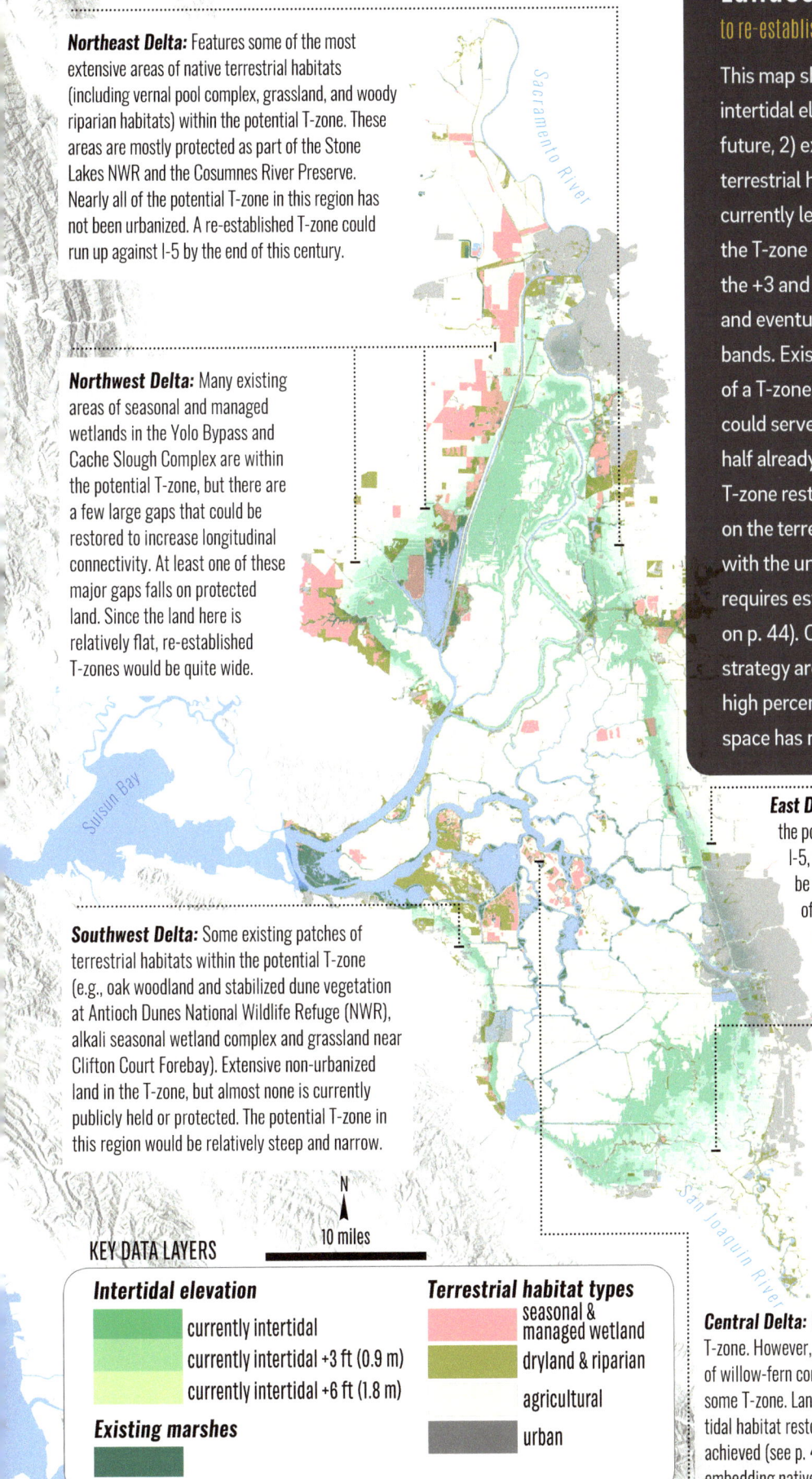

Northeast Delta: Features some of the most extensive areas of native terrestrial habitats (including vernal pool complex, grassland, and woody riparian habitats) within the potential T-zone. These areas are mostly protected as part of the Stone Lakes NWR and the Cosumnes River Preserve. Nearly all of the potential T-zone in this region has not been urbanized. A re-established T-zone could run up against I-5 by the end of this century.

Northwest Delta: Many existing areas of seasonal and managed wetlands in the Yolo Bypass and Cache Slough Complex are within the potential T-zone, but there are a few large gaps that could be restored to increase longitudinal connectivity. At least one of these major gaps falls on protected land. Since the land here is relatively flat, re-established T-zones would be quite wide.

Southwest Delta: Some existing patches of terrestrial habitats within the potential T-zone (e.g., oak woodland and stabilized dune vegetation at Antioch Dunes National Wildlife Refuge (NWR), alkali seasonal wetland complex and grassland near Clifton Court Forebay). Extensive non-urbanized land in the T-zone, but almost none is currently publicly held or protected. The potential T-zone in this region would be relatively steep and narrow.

Landscape considerations
to re-establish tidal-terrestrial transition zone processes

This map shows 1) areas that are currently at intertidal elevation or will be in the relatively near future, 2) existing tidal marshes, and 3) existing terrestrial habitats. Over the next century, if currently leveed areas are subject to tidal action, the T-zone would be expected to move through the +3 and +6 ft (0.9 and 1.8 m) elevation bands and eventually to the areas just upslope of these bands. Existing marshes can serve as the "bottom" of a T-zone, while exisiting terrestrial habitats could serve as the "top." Locations where either half already exists might be good for prioritizing T-zone restoration. The callouts here mostly focus on the terrestrial side of the tidal-terrestrial T-zone, with the understanding that a functional T-zone requires established tidal habitats (discussed on p. 44). Opportunities for implementing this strategy are bolstered by the fact that a relatively high percentage of the Delta's estuarine migration space has not been urbanized.[151]

East Delta: Relatively little urban development within the potential T-zone north of Stockton. The exception is I-5, which bisects the potential T-zone here and would be a barrier to continuous habitats and the movement of wildlife and materials. Some public/protected terrestrial habitats situated within the potential T-zone (e.g., Woodbridge Ecological Reserve and White Slough Wildlife Area).

South Delta: Some existing riparian habitats within the potential T-zone, but almost no other areas with native terrestrial vegetation. Some agricultural areas west of French Camp have the potential to contribute to the T-zone if restored. Historically much of the T-zone in the region was from tidal marsh to non-tidal marsh, an ecotone that could be re-established with proper habitat restoration and groundwater management. Almost no protected land.

Central Delta: Historically lacked any typical tidal-terrestrial T-zone. However, complexity within the tidal marshes (in the form of willow-fern complexes and sand mounds) would have created some T-zone. Land subsidence has constrained opportunities for tidal habitat restoration, but where subsidence reversal can be achieved (see p. 49), designs should consider the potential for embedding native terrestrial habitat types to create T-zones.

N

10 miles

KEY DATA LAYERS

Intertidal elevation

- currently intertidal
- currently intertidal +3 ft (0.9 m)
- currently intertidal +6 ft (1.8 m)

Existing marshes

Terrestrial habitat types

- seasonal & managed wetland
- dryland & riparian
- agricultural
- urban

Sacramento River

Suisun Bay

San Joaquin River

SFEI AQUATIC SCIENCE CENTER

PHYSICAL PROCESS GUIDELINES

1 **Allow tidal flows to access the transition zone (tidal hydrology)**

Since the upland extent of the T-zone is defined by the limits of tidal effects on the terrestrial environment, tidal flows are fundamental to a functional T-zone. The length and area of the T-zone increases significantly when tides flow over an intertidal marsh plain (and are not confined by levees). The fine silts and clays deposited by high tides form dense, nonporous soils, which contribute to the development of the unique habitat types found at the landward margin of the Delta's tidal marshes (e.g., alkali seasonal wetlands, wet meadows, and vernal pool complexes). High tides also contribute to important microhabitat heterogeneity in the T-zone through, for example, the formation of moisture gradients and wrack lines.

2 **Restore hydrologic connectivity from terrestrial to tidal wetland habitats (terrestrial hydrology)**

Although tidal hydrology is an important driver of the T-zone's distribution and extent, terrestrial hydrology also has a strong influence on 1) the composition of terrestrial habitat types[152] and 2) the transport of materials, organisms, and energy across the zone. Surface and subsurface flows draining from terrestrial areas to tidal marshes alter local hydrologic conditions and vegetation patterns, and deliver a variety of dissolved and suspended materials (e.g., nutrients, sediments, pollutants) to the marsh, which effects wildlife at both the individual and community levels.[153] Though humans can do little to control certain hydrologic variables that affect the T-zone (e.g., incipient rainfall), other important factors like groundwater levels, inundation frequency, and surface-runoff patterns can be managed. In general, we should improve hydrologic connectivity between terrestrial habitats and marshes through actions such as removing artificial barriers to surface flow across terrestrial habitats, replenishing groundwater levels in areas of the Delta's perimeter where they have been depleted, and allowing runoff to spread across the Delta's terrestrial habitats and into marshes (in the modern system, flood-control channels and agricultural ditches generally bypass both terrestrial and tidal environments and convey water directly to the Delta's interior channel network). Further research is needed, however, since relatively little is known about how the form of the T-zone influences the effects of terrestrial freshwater inputs into freshwater tidal wetlands.

3 **Allow for internal processes that maintain ecotones**

Though the T-zone is largely maintained by large-scale hydrologic processes that originate outside of the T-zone, there are a handful of small-scale internal processes that may contribute to T-zone persistence over time. Though not well-studied, positive internal feedback loops that could help maintain tidal-terrestrial ecotones in the Delta include prolific plant growth in freshwater marshes that excludes competing species; peat formation that alters the acidity of interstitial pore water and favors certain plant communities over others; and the establishment of fire-adapted shrub ecotones between freshwater marsh and oak woodland that are more likely to burn than the adjacent habitat types.[154]

4 **Allow space for migration/transgression of the transition zone upslope with sea-level rise**

As noted in the *Baylands Ecosystem Habitat Goals Science Update*, regional planning efforts must account for the likely landward migration of marshes and the T-zone due to sea-level rise. If the T-zone becomes compressed between rising waters and steep levees or urban development, its services will be diminished or lost completely. Therefore, broad areas for T-zone migration that can accommodate the full suite of local T-zone services in the future must be recognized as integral to the existing T-zone. The width of the T-zone that is needed for future accommodation space is dependent on sea-level rise, the slope of the areas involved, and the presence of built structures that could constrain the migration. It should be noted that, even with ample space, the landward migration of the T-zone is not a simple process—it is possible that migrating plant communities will encounter different soils and moisture regimes, which could drive a change in community composition.[155]

LANDSCAPE CONFIGURATION & SCALE GUIDELINES

5   The total length and longitudinal continuity of the transition zone should be maximized

Despite marsh fragmentation, which increases edge length, the total linear extent of the Delta's marsh-terrestrial transition zone has decreased by nearly 45% (from approximately 1,250 km to 700 km, as measured in *A Delta Transformed*).[156] What these numbers fail to convey is the severe fragmentation (and compression) of the T-zone. The average length of non-leveed T-zone segments in the Delta has decreased by more than 95%, from approximately ten thousand meters to a few hundred meters.[157] Long, continuous T-zones likely facilitate the migration and dispersal of plant and animal species, enabling them to move along the Delta between preferred patches of habitat.[158]

> **10,026 m** = mean **historical** T-zone segment length (SD = 17,154)[159]
>
> **364 m** = mean **modern** T-zone segment length (SD = 576; non-leveed)[160]

6   T-zones should be as wide as physically feasible

The width of the T-zone is not fixed. It varies over time and space based on the kinds and levels of ecosystem processes being considered, which means that no one T-zone width can be prescribed as a target or guideline.[161] T-zones will naturally be narrower in areas with low tidal range and steep slopes than in areas with a larger tidal range and gentler slopes. As a rule, in restored areas the width of contiguous marsh and terrestrial habitats on either side of their interface should be great enough to allow for the full range of T-zone ecosystem processes, and not artificially constrained by built infrastructure in order to support desired ecological functions. Semi-aquatic herpetofauna, for example, migrate from wetlands to adjacent terrestrial uplands for aestivating, basking, hibernating, feeding, and nesting. Generalized guidelines based on their movements dictate that core terrestrial zones should extend up at least 290 m beyond wetland boundaries.[162] Also important to consider are the distances terrestrial animals travel during daily or seasonal movements into the marsh to feed, which range from 20 m for California ground squirrels to more than a kilometer for herons and egrets (see table below). The T-zone should also be wide enough to accommodate expected sea-level rise and consequent habitat migration over the next century.

> **20 m** = distance California ground squirrels move from breeding habitats (oak woodland and grassland) into marshes to forage[163]
>
> **100 m** = distance California voles move into terrestrial habitats from marshes during wet season[164]
>
> **~100 m** = preferred distance between Tree Swallow nesting sites (e.g., oak woodland) and foraging sites (e.g., marsh)[165]
>
> **290 m** = terrestrial buffer that should be preserved upslope of wetlands to maintain terrestrial resources for herpetofauna[166]
>
> **1,000 m** = distance within which the amount of emergent wetland most strongly influences heron and egret colony site selection[167]

7   The diversity of transition zone types should be increased at the scale of the full Delta

In the historical Delta, marshes bordered a number of different terrestrial habitat types, which together created regional-scale diversity in the types of marsh-terrestrial T-zones. At least two types of T-zone have been entirely eliminated; though they once covered 39 and 45 linear kilometers respectively, transitions from marsh to stabilized interior dune scrub and from marsh to oak woodland/savanna are no longer found in the Delta. Since the suite of functions provided by each type of T-zone differs, restoration efforts should be coordinated across the Delta to ensure the full range of T-zone types are re-established, including those that may not currently be recognized as important due to their present-day absence. Analyses of how the extent, distribution, and composition of the Delta's marsh-terrestrial T-zone has changed can help guide this effort.[168]

Re-establish **connected terrestrial habitats around the periphery of the Delta**

In addition to their importance as part of the T-zone (pp. 66-69), the terrestrial habitats around the periphery of the Delta—including seasonal wetlands, such as wet meadow, vernal pool complex, and alkali seasonal wetland; dryland habitats, such as grassland, oak woodland/ savanna, and sand dunes; and woody riparian habitats, such as willow scrub and valley foothill riparian—were important for wildlife in their own right. The variety of terrestrial habitat types, which formed an unbroken ring around the Delta, augmented the overall biodiversity of the region and supported a large number of species that are now of special concern, including numerous endemic species.[169]

Currently, a majority of the Delta's terrestrial habitats are located in its interior, on artificial levees or in subsided areas formerly occupied by emergent freshwater wetlands. Though these areas are critical for supporting Delta wildlife in the near term, they are threatened by sea-level rise and levee failure over the long term. For this reason we focus on terrestrial habitats around the Delta's perimeter, which should be less vulnerable to these risks.

Though the large number of terrestrial habitat types precludes a detailed discussion of the specific processes that are important to each, edaphic and hydrological processes are influential across all of them. Specifically, there are a variety of important soil-forming processes—such as ferrolysis, clay and hardpan formation, and organic matter accumulation—that influence water movement, nutrient cycling, and species composition in the region's terrestrial habitats.[170] Hydrologic processes include surface inundation and groundwater-level fluctuation, which influence the distribution and composition of the Delta's terrestrial habitats.[171] Fire is also an important ecosystem process in some of the Delta's terrestrial habitats, especially oak woodland/ savanna.[172] Successful management and restoration of the Delta's terrestrial habitats will require a variety of tactics based on the site and relevant habitat types. These include native plantings, grazing management, fire management, and the reconfiguration of surface drainage patterns.

Image courtesy Google, Digital Globe

Terrestrial margin, Byron, CA. Near Clifton Court Forebay, remnant alkali seasonal wetlands sit above the Delta proper. These and other terrestrial habitats once formed an unbroken ring around the periphery of the tidal Delta, connecting the Delta to adjacent uplands and beyond. Though these critical peripheral Delta habitats are now diminished in size and heavily fragmented, there remains substantial opportunity for their re-establishment.

the Delta proper

alkali seasonal wetland

alkali seasonal wetland

upland annual grasslands

Northwest Delta: Historically a mixture of vernal pool complex, wet meadow, and seasonal wetland, as well as willow thickets. Remnant vernal pool complexes associated with Jepson Prairie Preserve are the largest contiguous tract of remnant terrestrial habitat types, and there are extensive seasonal wetlands further north. Land between these patches has not been urbanized, but is not public/protected. Restoration in these areas would improve the connectivity between existing patches of terrestrial habitats. There is some potential to restore woody riparian sinks at the distal ends of west-side tributaries, and to expand and enhance the protected terrestrial habitat corridor to Suisun Marsh.

Southwest Delta: Historically a mixture of alkali seasonal wetland, grassland, oak woodland, and stabilized interior dune vegetation. Interior dune vegetation could be restored near remnant dunes at Antioch Dunes NWR using remnant topography. Patches of remnant alkali seasonal wetland complex are proximal to vast protected areas higher in the watershed. Targeted land acquisition and restoration could complete a continuous corridor of natural habitats from the hills to the Delta.

Landscape considerations
to re-establish connected terrestrial habitats around the periphery of the Delta

This map shows the terrestrial habitats around the periphery of the Delta. The map highlights areas above the approximate future elevation of MHHW with 6 ft (1.8 m) of SLR by adding a transparency to the area below this elevation (making it lighter in color). In general, the higher areas should be less vulnerable to continued SLR and levee failure. Most of the periphery has not not been urbanized, which means there are opportunities to increase the extent, diversity, and connectivity of the native terrestrial habitat types found along the Delta's edge.

Northeast Delta: Historically a complex mosaic of non-tidal marsh, woody riparian habitat (associated with the Cosumnes Sink and Mokelumne River), grassland, vernal pool complex, wet meadow, and seasonal wetland. There are terrestrial habitats associated with the Cosumnes River Preserve and Stone Lakes NWR, but there is a notable gap in terrestrial habitats between these two preserves. Numerous protected parcels in this area could be restored to increase their connectivity. Expanding terrestrial habitats here will be key, since substantial portions of the existing terrestrial habitats could become tidal before 2100.

South Delta: Historically supported non-tidal marsh and woody riparian habitats in the extensive San Joaquin River floodplain. Beyond the floodplain was mostly alkali seasonal wetlands to the east, with wet meadow/seasonal wetlands and grasslands to the west. Some woody riparian habitats remain, but no other substantial terrestrial habitat patches. There is substantial agricultural land with restoration potential, but very little is public/protected and most has a high risk of urbanization.

East Delta: Historically featured oak woodlands and alkali seasonal wetlands. There are opportunities to restore alkali seasonal wetlands where remnant Devries Sandy Loam soils (slightly alkaline, somewhat poorly drained) still exist. Remnant oak woodlands at Oak Grove Regional Park in Stockton could serve as a seed source for restoring oak woodlands in natural areas.

KEY DATA LAYERS

Elevation

- future MHHW contour line (current + 6 ft [1.8 m])
- future supratidal
- future tidal

N

10 miles

Terrestrial habitat types

- willow thicket
- riparian scrub & shrub
- valley foothill riparian
- wet meadow & seasonal wetland
- vernal pool complex
- alkali seasonal wetland
- stabilized interior dune vegetation
- grassland
- urban

SFEI AQUATIC SCIENCE CENTER

PHYSICAL PROCESS GUIDELINES

1 **Re-establish areas of short-duration inundation along the Delta's margin (surface hydrology)**

In the historical system, most of the Delta's seasonal wetlands were fed by small intermittent or ephemeral streams that emanated from the surrounding foothills, lost channel definition before reaching the wetlands, and temporarily or seasonally inundated the land. These terrestrial areas often featured natural topographic variability (e.g., hog-wallows, swales, channel ridges), which further contributed to the development of moisture gradients and habitat heterogeneity.[173] Today, an expansive network of drainage ditches and flood-control channels has altered the regional surface hydrology and severely reduced the extent of short-duration inundation in terrestrial areas along the Delta's margin. Where possible, efforts should be made to re-establish historical drainage patterns and to increase the extent of land along the Delta periphery subject to beneficial flooding.

2 **Maintain groundwater conditions for the full suite of terrestrial habitat types (subsurface hydrology)**

Groundwater levels and flows influence the Delta's terrestrial habitats at a number of different scales. At the regional scale, the availability and depth to groundwater influences the distribution of plant species and the habitat types they form. At a finer scale, natural patterns in groundwater movement help maintain local heterogeneity in the water chemistry of wetland features.[174] Localized groundwater depressions from municipal pumping have likely altered the ability of certain areas to support certain terrestrial habitats. Large decreases in the regional water table level would, for example, be expected to negatively impact alkali seasonal wetlands and oak woodlands.[175] On the other hand, such decreases would not be expected to affect vernal pool habitats, which are influenced instead by the local perched aquifer. Perched groundwater discharged to vernal pools stabilizes the vernal pool water levels (increasing their inundation duration and extent) and buffers the chemistry of vernal pool water, which influences the abundance and composition of vernal pool flora and fauna. Though not sensitive to the regional water table level, shallow perched aquifers and the habitats that rely on them are likely sensitive to even small changes in local land use, which should be considered in regional planing efforts.[176] Because areas with high groundwater can provide climate refugia for terrestrial species, maintaining groundwater levels is a priority for buffering terrestrial ecosystems against climate change.[177]

3 **Identify remnant soil types and re-establish edaphic processes that support terrestrial habtiats**

Historically, soil properties were one of the primary factors determining the regional distribution of terrestrial habitat types. Seasonal wetlands, for example, were supported on clay soils, grasslands and oak woodlands favored well-drained loams, and interior dune scrub occupied the sandy mounds of Eastern Contra Costa County.[178] Remnant soil types should be located to identify areas where efforts to restore historical terrestrial habitats are most likely to succeed. On the eastern margin of the Delta, for instance, at the former sites of large alkali seasonal wetlands, the soils remain slightly alkaline and somewhat poorly drained, suggesting this could be a good place to re-establish historical habitats in their historical location. In other areas, the re-establishment of the historical habitat type might be precluded by land-use practices that have altered soil conditions (such as the cutting and filling of mounds and depressions to level land, excavation of drainage ditches to lower shallow water tables, and deep ripping to increase permeability of subsoil horizons).[179] These areas might favor the restoration of another terrestrial habitat type, or first require the re-establishment of key edaphic processes and soil structures. Whether such actions are feasible should be evaluated and addressed during the planning phases of new projects. In all cases, planners and managers should consider and work to re-establish the processes that support the development and maintenance of desired soil conditions over time.

LANDSCAPE CONFIGURATION & SCALE GUIDELINES

4   Maximize the patch size of terrestrial habitat types at the periphery of the Delta

The patch size of the Delta's terrestrial habitat types should be increased, since, in general, the ecosystem functions desired from terrestrial habitats increase with patch size (see the tables below for examples).

Vernal pool complex	**37 ha** = area to protect Western Pond Turtle overwintering (centered around a pool)[180]
	1,375 ha = area to protect breeding population of California tiger salamanders (centered around a breeding pool)[181]

Wet meadow / seasonal wetland	**129 ha** = minimum recommended giant garter snake habitat patch size[182]

Oak woodland	**39–2,878 ha** = mule deer home range[183]

Grassland	**8 ha** = mean White-tailed Kite hunting territory size[184]
	40 ha = size below which Swainson's Hawk foraging suitability decreases[185]
	160 ha = badger home-range size[186]
	189 ha = mean Western Burrowing Owl home-range size[187]
	336 ha = minimum Swainson's Hawk home-range size[188]

5   Maximize the width of terrestrial habitats buffering seasonal wetlands

Just as tidal wetlands should have continuous transitions to upslope terrestrial habitats (p. 66), so too should the seasonal wetlands along the Delta's margins. Some of the clearest research on the necessary widths of terrestrial buffer habitats around seasonal wetlands comes from research on the California tiger salamander. Like many other species of herpetofauna, the salamanders are semi-aquatic organisms that make migrations from wetlands into adjacent terrestrial habitats for a portion of their life-cycle. As noted on page 69, generalized guidelines for breeding herpetofauna dictate that core terrestrial zones should extend up at least 290 m beyond wetland boundaries.[189] However, for vernal pool complexes used by tiger salamanders, associated terrestrial habitat zones should be even larger, since the salamanders are capable of migrating nearly 2.5 km from breeding ponds into surrounding terrestrial habitats (grassland and oak woodland/savannas).[190]

6   The continuity of high-quality terrestrial habitats within the Delta should be maximized

The continuity of terrestrial habitats in the Delta should be maximized to facilitate the unimpeded movements of native wildlife. This applies both to terrestrial habitats along the periphery of the Delta, and to the terrestrial riparian habitats that extend into the tidal zone along natural levees (see pp. 60-65).

7   Connectivity to terrestrial areas outside of the Delta should be increased

On its eastern margin, the tidal Delta lies only 25 km (15 mi) from the extensive Sierra Nevada foothill woodlands, and 65 km (40 mi) from montane forests. The Cache Slough Complex lies only 8 km (5 mi) from Suisun Marsh. Parts of the South Delta are less than 2 km (1 mi) from the western margin of the Coast Range. Large-scale connectivity between these areas outside of the Delta should be increased through the creation of protected corridors (especially along streams that connect different parts of the watershed) and through other methods that increase landscape permeability to native wildlife movement.[191]

Expand **wildlife-friendly agriculture**

SUPPORTED FUNCTIONS

marsh wildlife
(see pp. 88-91)

edge wildlife
(see pp. 100-103)

waterbirds
(see pp. 96-99)

fish
(see pp. 84-87)

productivity
(see pp. 104-105)

Expanding wildlife-friendly agriculture and continuing to integrate agricultural lands into conservation planning within the Delta is a critical strategy for working landscapes that support both people and wildlife.[192] Though agricultural lands are not a substitute for natural habitat types, if managed properly they can be utilized by many native species and help mitigate the loss of natural habitats in areas where agriculture dominates.[193] In the Central Valley (including the Delta), agricultural lands provide critical support for waterbirds, particularly wintering waterfowl and Sandhill Crane.[194] Agricultural fields are critical for supporting other species as well, including anadromous fish, Tricolored Blackbird, and Swainson's Hawk.[195] In addition to physical disturbances to wildlife when crops are planted or harvested, agricultural practices both within and outside of the Delta also have the potential to impact Delta wildlife through water diversions and effects on water quality, so actions within the Delta should be coordinated with state and regional planning efforts.

Supporting wildlife on agricultural lands can support tourism, hunting, and other recreational activities, and provide mutual benefits for agriculture. Where insectivorous bats are abundant, they are known to make significant impacts on agricultural insect pests.[196] Terrestrial habitats adjacent to agricultural lands also support populations of wild crop pollinators, chiefly bee species, which can provide tremendous value to farmers.[197] Flooding rice fields in the winter to promote the decomposition of rice straw is mutually beneficial for waterfowl and crop yields.[198] Incentive programs included in the Farm Bill can make wildlife-friendly farming a more viable option for farmers.[199]

In addition to the many best-management practices, specific tactics for implementing this strategy might include creating seasonal or permanent wetlands within fields (including so-called "pop up wetlands"),[200] flooding fields to mimic floodplain processes,[201] planting hedgerows and buffer strips,[202] "re-oaking" the agricultural landscape,[203] adjusting scheduled fieldwork, reducing pesticide and herbicide application, and the implementation of water-conservation measures.

Image courtesy Google

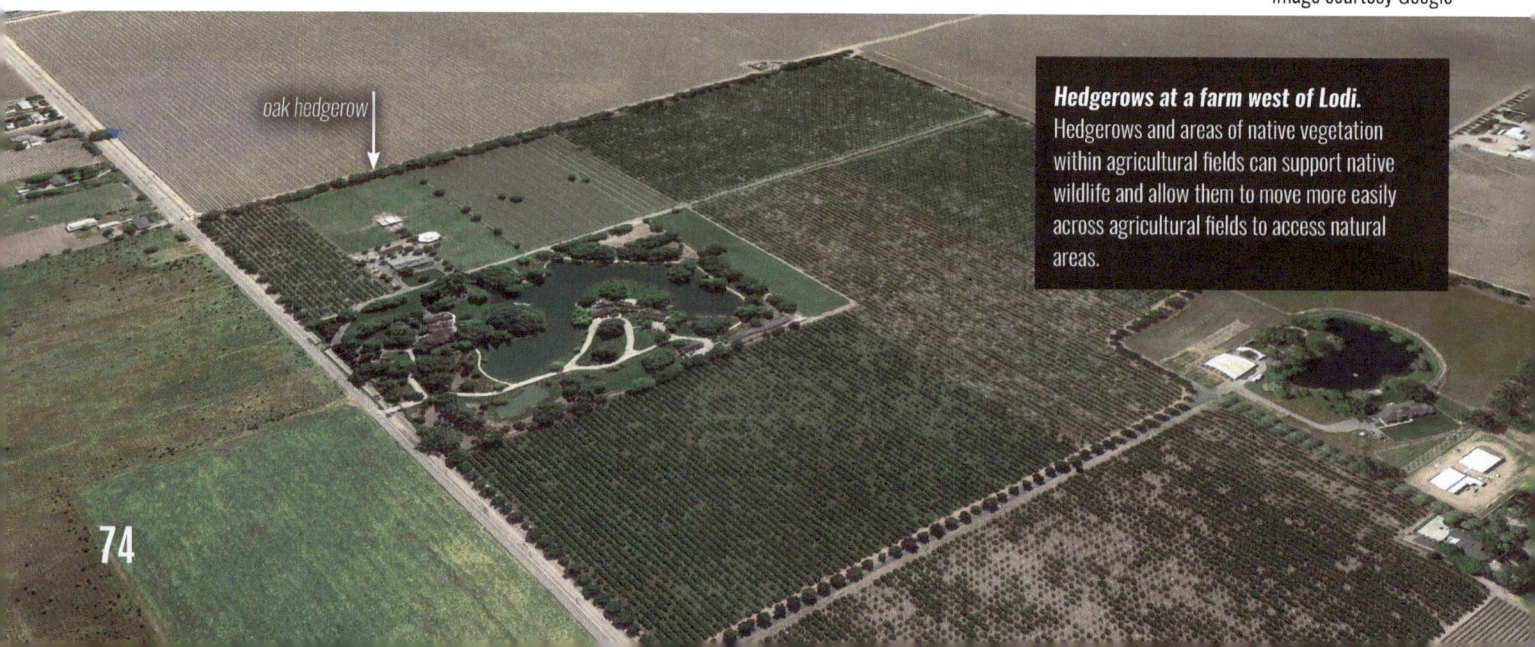

oak hedgerow

Hedgerows at a farm west of Lodi.
Hedgerows and areas of native vegetation within agricultural fields can support native wildlife and allow them to move more easily across agricultural fields to access natural areas.

Landscape considerations
to expand wildlife-friendly agriculture

This map shows agricultural lands in the Delta. Integration of wildlife-friendly farming with conservation planning should consider the proximity to natural areas that support similar species. Supratidal areas may be more sustainable locations for wildlife-friendly agriculture over the long term, given concerns about sea-level rise, subsidence, and levee failure. Two examples of successful wildlife-friendly agricultural projects are highlighted here; however, other ongoing efforts and further opportunities to improve support for wildlife on agricultural lands exist throughout the Delta.

Nigiri Project: Rice fields at Knaggs Ranch in the Yolo Bypass are flooded to provide rearing habitat for juvenile Chinook salmon. Salmon grow quickly in these highly productive fields, increasing their likelihood of surviving migration.[204]

Conaway Ranch: Several conservation easements exemplify the success of different wildlife-friendly agriculture approaches. A 224-acre easement comprised of high-value wetland (cattails, bulrush) and upland (milk thistle and mustard) nesting habitat supports large breeding colonies of Tricolored Blackbirds. A thousand-acre easement of rice lands with open-water channels and patches of emergent vegetation supports the giant garter snake. These examples, plus several other voluntary wildlife-friendly practices, have helped support native species for many years, while the ranch also balances an effective farming operations.[205]

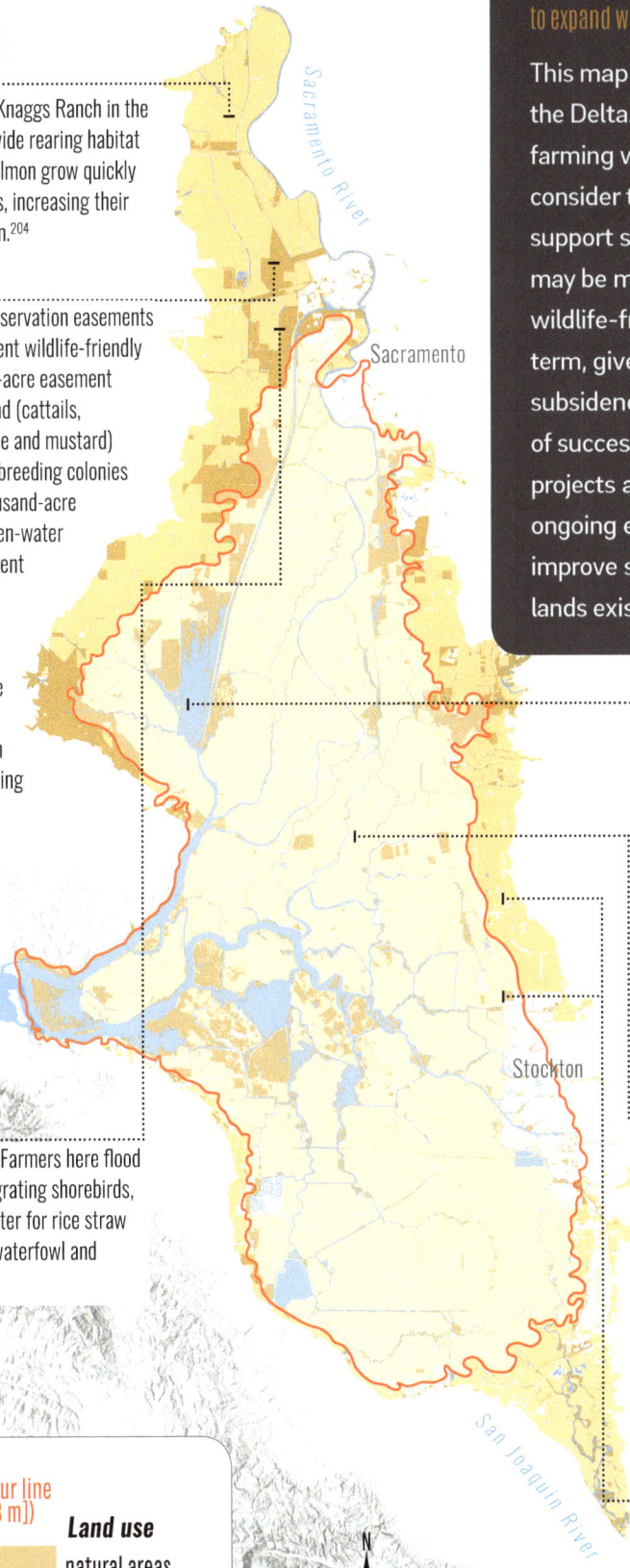

Experimental flows: State agencies partnered with Sacramento Valley farmers to create a fall plankton bloom to support Delta smelt. In July 2016, agricultural districts began pumping 15,000 acre-feet of Sacramento River water into irrigation canals, through the Yolo Bypass, and into the North Delta to mimic historical floodplain conditions. This effort successfully created a noticeable spike in phytoplankton, paving the way for more such experiments and partnerships.[207]

Staten Island: Staten Island provides critical support for Sandhill Cranes, as well as other waterbirds, including Snow Geese and Tundra Swans. Wintering waterfowl are allowed to forage uninterrupted in grain crops and flooded fields, and have little impact on annual crop yields. Staten Island's proximity to the Cosumnes Preserve, which also supports Sandhill Cranes, increases the area's value in supporting these birds.

Yolo Bypass Wildlife Area: Farmers here flood fallow fields in summer for migrating shorebirds, and flood rice fields during winter for rice straw decomposition and to benefit waterfowl and wintering shorebirds.[206]

Re-oaking: Remnant oak woodlands at Oak Grove Regional Park in Stockton could serve as a seed source for re-oaking the adjacent agricultural landscapes, especially in hedgerows. Some farms to the north of the park have extensive hedgerows and small oak woodlands embedded in the agricultural matrix.

KEY DATA LAYERS

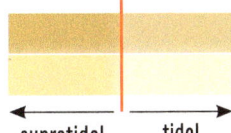

future MHHW contour line
(current + 6 ft [1.8 m])

Land use
natural areas
agriculture

⟵ supratidal   tidal ⟶
Future elevation

N

10 miles

PHYSICAL PROCESS GUIDELINES

1 **Minimize detrimental impacts to water quality from agriculture**

Application of pesticides, herbicides, and fertilizers to agricultural fields can impact water quality in the Delta. Concentrations of some pesticides are high enough to elicit concern about impacts to wildlife.[208] Impacts to water quality might be minimized through best management practices, including reduced pesticide use, integrated pest management, settling basins, and buffer strips to filter runoff.[209]

2 **Minimize detrimental impacts of water diversions for agriculture**

Large-scale water diversions, such as the Central Valley Project and the State Water Project, substantially impact the freshwater flows critical to supporting native fish.[210] Water export for agricultural and municipal demand have led to reverse flows along the Old and Middle rivers of the San Joaquin, potentially increasing entrainment of native fish.[211] Water conservation measures throughout the state can reduce Delta exports. Within the Delta, intakes and outfalls affect water quality and flows. Fish screens can prevent direct mortality due to fish entrainment.[212]

Sandhill Cranes at Staten Island, 2013, photograph courtesy Ken Phenicie Jr.

3 **Support diverse and dynamic wildlife habitats via flexible and responsive management in agricultural areas**

Different crops have potential to provide support for different wildlife species. Row crops and rice fields may be more amenable to supporting waterbirds and fish, while hedgerows to support movement of terrestrial wildlife may be easiest to maintain on vineyards and orchards.[213] Increasing conversion of agricultural lands from row crops to vineyards and orchards may reduce the availability of lands that can support wetland species. For some highly mobile species, frequent monitoring and responsive managing can funnel resources to where needs are greatest based on real-time wildlife use. For example, The Nature Conservancy's "pop-up habitats" divert water to farms when waterbird densities are high.[214]

LANDSCAPE CONFIGURATION & SCALE GUIDELINES

4 **Increase the acreage of agricultural fields managed in a way that mimics seasonal wetland or floodplain habitat**

Some wildlife-friendly agricultural practices in the Delta create seasonal or perennial wetlands that mimic the hydrology of historical wetlands. Shallow flooded fields that provide roosting habitat for Sandhill Cranes and shorebirds have vegetative structure and flooding depths similar to the seasonal wetlands that have been largely lost along the periphery of the Delta.[215] Rice fields provide long-duration floods and invertebrate-rich rearing habitats, which flood basins provided historically. These agricultural wetlands can support high densities of wintering and migrating waterbirds as well as fish, and are critical to supporting these species in the absence of extensive natural wetlands. Agricultural wetlands support different species depending on crop type, flooding patterns, and post-harvest practices.[216]

5 **Increase support for native terrestrial wildlife through hedgerows and native vegetation**

Patches of native vegetation within or between agricultural fields, whether remnants of historical habitats (e.g., oak trees, vernal pools) or linear features along the edge of fields (e.g., buffer strips, hedgerows), can provide habitat for native wildlife and easier movement through the landscape. Studies of hedgerows in Europe found **hedge size (height/width/volume) and the presence/abundance of trees were both positively correlated with the species richness and abundance of breeding birds using hedgerows.**[217] However, the use of hedgerows as a tool to increasing the long-term viability of wildlife populations is equivocal.[218] Increasing landscape connectivity via hedgerows and buffer strips also carries the risk of aiding the dispersal of nuisance species (e.g., Norway rats, feral cats) at the expense of native species, and more research is needed to understand the best management practices for the Delta.

6 **Minimize the distance from wildlife-friendly agriculture to nearby wildland areas that benefit similar species**

Species supported by wildlife-friendly agriculture can benefit from the close proximity of appropriate wildland habitats.[219] Understanding how landscape configurations and distances between agricultural habitats and wildland habitats impact particular species will help with more integrated landscape planning in the future. More research into this subject is needed.

Integrate **ecological functions into urban areas**

SUPPORTED FUNCTIONS

marsh wildlife
(see pp. 88-91)

edge wildlife
(see pp. 100-103)

riparian wildlife
(see pp. 92-95)

Integrating support for native wildlife into urban areas can provide supplementary habitat for certain species and can help connect people to nature. Urban areas within the modern Delta are primarily built on high ground, in areas that were riparian forests, oak savannas or seasonal wetlands historically. Wildlife adapted to these historical habitats may be supported in urban open spaces through the preservation of remnant natural areas, the planting of native species, or through vegetation and hydrology that mimic the structure of natural systems in urban spaces.

There is still much to learn about how urban areas can support wildlife, and such actions should be implemented carefully and monitored to prevent inadvertently increasing support for pest species or human-wildlife conflict. Despite these uncertainties, efforts to increase support for native wildlife of the region can be important for educating and engaging the public, fostering a sense of place, and potentially increasing local wildlife population sizes. Urban green spaces that support wildlife can provide additional benefits to people, such as purifying air and water, moderating local climate, reducing noise pollution, increasing real-estate values, improving neighborhood and landscape aesthetics, and enhancing psychological well-being.[220]

Possible tactics for implementing this strategy include urban greening and native plantings (such as "re-oaking"),[221] low-impact development,[222] beneficial reuse of wastewater, and construction of water-treatment wetlands.

Image courtesy Google

City of Davis

Davis wastewater treatment wetlands

Davis watewater treatment wetlands.
Wastewater treatment wetlands, like those that treat water from the City of Davis, remove aquatic pollutants through a variety of biological, physical, and chemical processes. These constructed wetlands generally operate at a relatively low cost and offer a variety of co-benefits to people and native wildlife.[223]

Landscape considerations
to integrate ecological functions into urban areas

Opportunities to increase support for native wildlife exist in the towns and cities around the Delta. Measures that increase open space, native vegetation, and controls on urban stressors (like non-native predators and road-kill mortality) can provide benefits to wildlife and people. Actions taken in large population centers may have the most potential to connect the greatest number of people to native wildlife.

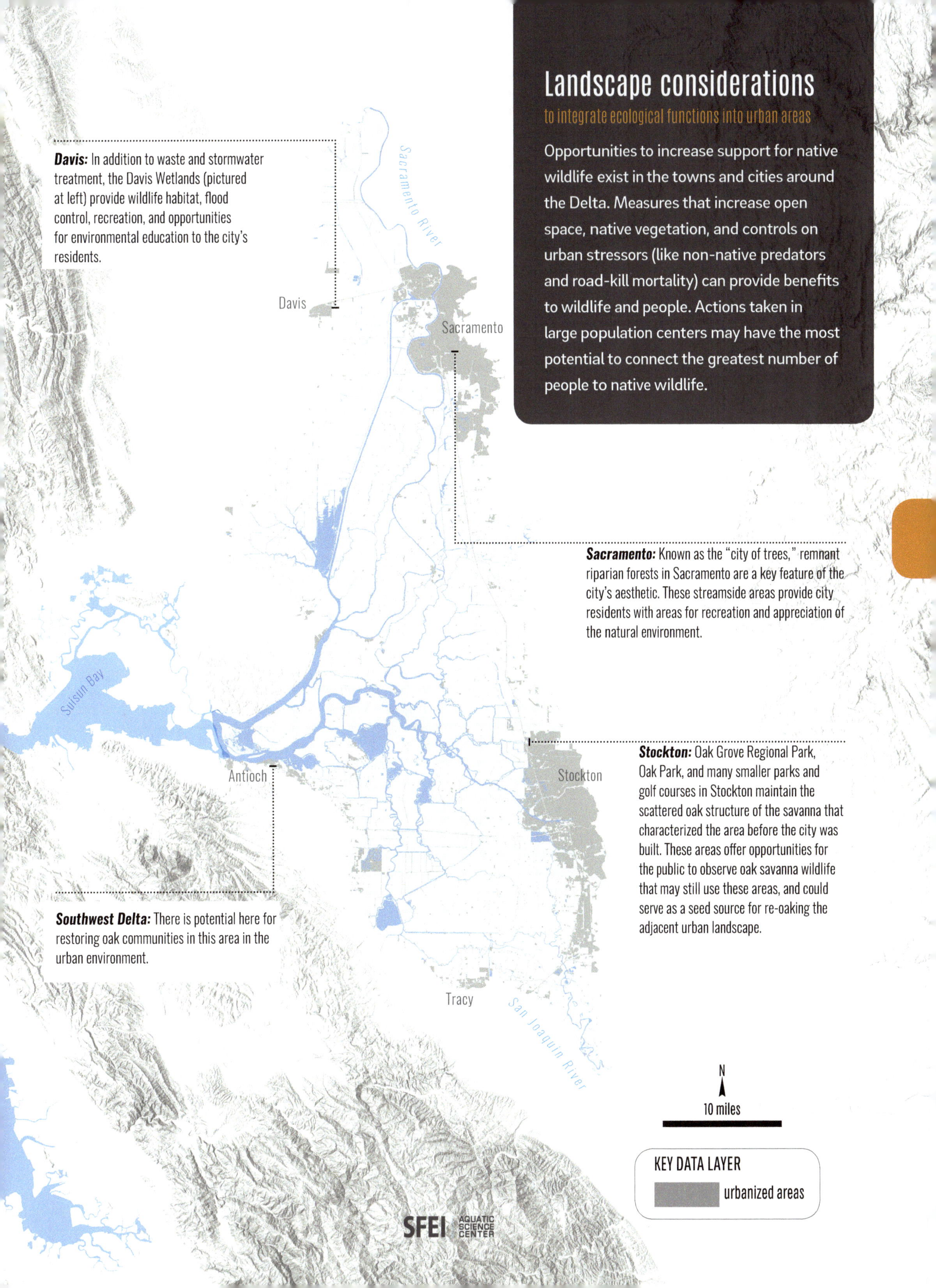

Davis: In addition to waste and stormwater treatment, the Davis Wetlands (pictured at left) provide wildlife habitat, flood control, recreation, and opportunities for environmental education to the city's residents.

Davis

Sacramento

Sacramento: Known as the "city of trees," remnant riparian forests in Sacramento are a key feature of the city's aesthetic. These streamside areas provide city residents with areas for recreation and appreciation of the natural environment.

Suisun Bay

Sacramento River

Antioch

Stockton

Stockton: Oak Grove Regional Park, Oak Park, and many smaller parks and golf courses in Stockton maintain the scattered oak structure of the savanna that characterized the area before the city was built. These areas offer opportunities for the public to observe oak savanna wildlife that may still use these areas, and could serve as a seed source for re-oaking the adjacent urban landscape.

Southwest Delta: There is potential here for restoring oak communities in this area in the urban environment.

Tracy

San Joaquin River

N

10 miles

KEY DATA LAYER

urbanized areas

PHYSICAL PROCESS GUIDELINES

1 | **Minimize impacts to water quality from urban areas**

Cities and towns in and around the Delta have the potential to impact water quality through runoff and wastewater discharges. Pollutants and other stressors from cities that have the potential to impact water quality include heavy metals, PCBs, excess nutrients, bacteria, and pharmaceutical and personal-care products.[224] Treatment of wastewater and storm-water runoff, buffers around wetland and aquatic habitats, and reducing contaminant inputs can lessen these impacts. The re-design of the Sacramento Municipal Wastewater Treatment plant to reduce nitrogen and ammonia is a large-scale experiment in nutrient removal. Stockton's aeration facility was created to address a hypoxic zone created in the San Joaquin River.

2 | **Restore appropriate fluvial zone processes in urban creeks**

Urban creeks and associated riparian zones can function as wildlife corridors through highly developed areas, allowing critical passage for both terrestrial and aquatic animals. Maintaining healthy urban streams means managing the local watershed to allow adequate flows and transport of sediment and other materials. This includes actions such as buffering urban streams, allowing space for wide riparian zones to establish, implementing low impact development, and reducing the extent of impervious surfaces. Some of these actions also have the potential to provide flood-protection benefits for people if implemented correctly.

LANDSCAPE CONFIGURATION & SCALE GUIDELINES

3 **Increase native wildlife quality and quantity of habitat in urban areas**

Support for wildlife in urban areas can be improved by increasing the size and amount of open space in and around cities. Urban open spaces not only include city parks but also golf courses, cemeteries, green roofs, transportation right of ways, and utility corridors. While larger open-space areas often support greater numbers of species,[225] even small areas can serve critical ecological functions. Remnant isolated patches as small as 50–1,000 m² have proven useful for sustaining populations of invertebrates and can serve as stepping-stone habitats.[226] Native species can also be supported by planting native vegetation in yards and as street trees.

Many of the areas that support cities and towns today supported terrestrial and tidal-terrestrial T-zone habitats historically. [227] Remnants of the former landscape and restoration of native plant communities and habitat structure may be able to support some of the species adapted to these areas, particularly highly mobile species like birds, bats, butterflies, and other insects. While many species are capable of inhabiting urban areas (occasionally even rare and endangered species), not all species have the potential to thrive in these areas.[228]

4 **Mimimize impacts of urban stressors and hazards to wildlife**

While urban areas have the potential to support wildlife, they also present hazards for the species that live there. Risks to wildlife include increased mortality from collisions with cars, power lines, and buildings, and from predation by urban-exploiting species, such as Norway rats. There are also sublethal stressors, including air, light, and noise pollution (in addition to the issues with water quality described earlier). Measures such as wildlife passages over or under roads, bird-safe building designs, keeping home cats indoors, and control of pest species can help minimize these risks.

5 **Increase opportunities for people to connect with nature**

Increased contact with nature and increased exposure to green spaces has benefits for human health and well-being.[229] In addition, knowledge of and connection to local ecosystems and wildlife is critical to maintaining support for conservation measures. Restoring native plant and animal communities within urban areas can encourage people to go outside, reap the benefits of time in nature, and learn about local ecosystems.

Near Isleton, 2012, photograph courtesy Michele Ursino

SUPPORTING ECOLOGICAL FUNCTIONS IN THE FUTURE DELTA

American Bittern, San Joaquin River, photograph courtesy Becky Matsubara

In this section we summarize how the recommendations and strategies presented earlier in this report could work together to support the resilience of desired ecological functions in the future Delta—recovering lost support for native wildlife and helping those species to persist in a changing environment. In this chapter we focus on the ecological functions identified in our previous companion report, "A Delta Transformed."[1] These ecological functions were chosen to reflect landscape-scale support for native species at both the population and community levels. The majority of these functions focus on life-history support for particular wildlife groups, including marsh wildlife, native fish, riparian wildlife, waterbirds, and wildlife inhabiting the terrestrial perimeter of the Delta. In addition, we discuss food-web support, with an emphasis on ongoing research around primary production that has developed as an offshoot of this project. Finally, we discuss recommendations for supporting native biodiversity in the future Delta, and how our overall focus on ecological function both supports and complements a focus on biodiversity. Further work is needed to assess the constraints to successful implementation of these recommendations, and how such impediments might change over time.

ECOLOGICAL FUNCTIONS

Supporting **native fish** in the future Delta

Supporting **marsh wildlife** in the future Delta

Supporting **riparian wildlife** in the future Delta

Supporting **waterbirds** in the future Delta

Supporting **edge wildlife** in the future Delta

Supporting **productivity** in the future Delta

Supporting **biodiversity** in the future Delta

Supporting **native fish** in the future Delta

The Delta historically supported a diverse fish community that included the endemic delta smelt, multiple large runs of Chinook salmon, and numerous other estuarine and anadromous fish.[2] The fish community in the Delta has undergone a dramatic transformation,[3] mostly as a result of large-scale habitat change, flow alterations, and the introduction of non-native species. Habitats that supported native fish historically were dynamic and complex, providing many areas of slow moving water and abundant food resources.[4] These habitat types included floodplains, dendritic marsh channels, riparian forests along streams, ponds, and lakes. Large losses of wetlands (including floodplains), disconnection of wetlands from open water, and homogenization of channels have led to far fewer resources for native fish and dramatic declines in the populations of many species. Invasive SAV and FAV have also altered the structure of habitats available to native fish.[5] Both habitat change and invasive species (e.g., *Corbicula* clams) have limited the food resources available to native fish.[6] Partly as a result of these changes, the Delta has become a novel ecosystem that supports many non-native fish species.[7] While we focus on increasing support for native fish in the future Delta, actions taken to benefit native fish are likely to also benefit non-native fish,[8] including important sport fish such as striped bass, and non-native fish that are important prey species for other wildlife.

Supporting native fish in the Delta over the long term will mean recreating a complex and variable aquatic landscape with extensive low-velocity, high-productivity aquatic and wetland habitat types. Recreating these conditions will require naturalistic flow regimes, which benefit native fish over non-natives,[9] and adequate sediment supplies to maintain habitats. Anadromous fish need multiple routes of unimpeded passage through the Delta with numerous places for foraging and refuge. For some native fish, further research into specific habitat factors limiting the population (e.g., spawning habitat needs) is needed. It may not be possible to restore native dominance in many parts of the Delta. At least in the short term, intensive effort may be required to control invasive species and manage water to create areas where natives can thrive, and some areas may have to be prioritized over others. Climate change will exacerbate stressors on native fish by further altering the hydrologic patterns and water quality.[10] Of particular concern is the projected increase in water temperature that will impact species already near the limit of their physiological tolerance.[11]

Support for native fish can be better integrated with other land uses, such as agriculture and flood control, by managing floodplains for multiple benefits (e.g., Yolo Bypass), improving water quality in run-off, and reusing wastewater to benefit the ecosystem. In addition, the health of the Delta fish community is vital for supporting subsistence fishing, sport fishing, and hatchery salmon.

Delta smelt, Livingston Stone Fish Hatchery, photograph courtesy Steve Martarano (USFWS)

Recommendations

- *Restore functional flows* of sufficient magnitude and appropriate timing to activate floodplains, provide the physical and chemical cues needed for migration, and create tolerable environmental conditions (e.g., temperature, turbidity) for native fish. This includes reducing reverse flows and the risk of entrainment. See "fluvial zone processes", pp. 60-65.

- *Create a diversity of wetland and aquatic habitat types,* re-establishing landscape-scale habitat type configurations that emulate historical patterns where appropriate and achievable. Focus particularly on high-productivity, low-velocity, intermittently flooded habitat types, and create seasonal and interannual variability within these habitat types).

  - **Restore complex floodplains and intermittently flooded basins** that provide opportunities for foraging and spawning, as well as refuge from predators and physical stressors. Floodplains should mimic natural flooding patterns and remain flooded for long enough to activate food webs and support fish rearing and spawning. The unique flood basin habitats of the North Delta have been lost. Manage remaining lakes (e.g., Stone Lakes) as intermittently flooded habitats to give advantage to native fish over non-natives.

  - **Restore multiple large marshes** with complex dendritic channel networks to support juvenile salmon rearing and marsh fish, and to diversify and increase primary production available in aquatic habitat types.

  - **Restore complex channel networks** that provide spatial and temporal habitat heterogeneity. Channel geometries should relate to tidal excursion to achieve variation in residence time, and channels should vary in water speeds, depth, light, and turbidity.

  - **Restore riparian areas** that provide important food resources for fish, add to bank stability and channel heterogeneity, and provide shade to maintain cooler water temperatures.

- *Choose landscape configurations that provide resources and refuges for native fish throughout the Delta.* Maintain marshes, floodplains, and other off-channel habitats, as well as complex channel networks, across different regions of the Delta to ensure habitat availability along physical gradients (e.g., salinity and temperature).

- *Maintain connectivity between high-value habitat types within the Delta.* Maintain flows and bathymetry to support movement of native fish between areas with high habitat value. Increase the habitat value of levees and channel edges to support landscape connectivity for native fish.

- *Maintain appropriate connectivity between wetlands and open water* to allow exchange of energy, materials, and organisms between wetlands and adjacent aquatic habitat types. Connectivity with wetlands has the potential to improve water quality, provide food resources, and impact turbidity in open water areas.

- *Expand fish-friendly management practices* such as fish-friendly farming, grazing, ponds, and manage wetlands to provide additional foraging habitat for fish, reduce the impacts of water diversions, and produce effluent that is cooler and has fewer contaminants.

- *Increase cool water conditions in the Delta.* While the extent to which this can be achieved is not well understood, possible interventions include maintaining cool groundwater inputs to the Delta, restoring marshes to allow evaporative cooling on the marsh plain, maintaining riparian cover to shade channels, and managing upstream dams to release cold-water pulses.

Short-term to long-term planning

In the short term, the Northwest Delta may provide the greatest opportunity to increase support for native fish (i.e., the "Yolo Bypass-North Delta-Suisun Arc"). However, over the long term, areas that have less potential to support fish today may become more valuable as water management and land-use change. Sufficient sediment supplies to the Delta will be critical to support the persistence of marshes and floodplains, and for optimal turbidity. Interim habitats created in the early stages of tidal marsh restoration projects can create temporary habitats for fish. Further research is needed to understand the potential support these interim habitats can provide, and how these benefits can be optimized.

How strategies fit together to support native fish

Creating the complex, appropriately connected habitat types necessary to support native fish will require restoring tidal, fluvial, and terrestrial transition-zone processes to maintain a diverse mosaic of complementary habitat types. Wildlife-friendly agriculture has the potential to increase support for fish, and integrating wildlife-friendly agriculture into landscape-scale planning could maximize such benefits.

Re-establishing tidal zone processes in channels and flooded islands provides

- phytoplankton production
- flow variability that creates velocity refugia and supports fish movement
- complex channel networks that provide refuge from predators
- additional habitat complexity, cover, and food in SAV/FAV; though this particularly supports non-native fish

Integrating ecological functions into urban areas provides

- reduced runoff of chemicals that negatively impact fish and prey
- beneficial reuse of treated wastewater

Re-establishing connections between streams and tidal floodplains provides

- more dynamic, complex habitats

Re-establishing tidal-terrestrial transition zone processes provides

- buffer habitat to keep upland and urban stressors from impacting wetland and aquatic habitats that support fish

Re-establishing tidal marsh processes in areas at intertidal elevations provides

- marsh-derived food
- improved water quality, temperature, turbidity
- rearing habitat for salmonids and foraging habitat for littoral fish

Expanding wildlife-friendly agriculture provides

- reduced runoff of pesticides that negatively impact fish and prey resources
- less freshwater diversions to progress towards net positive flows
- fish access to fields as rearing habitat

Re-establishing fluvial processes along streams provides

- delivery of freshwater from the watershed
- proper cues for migration
- sediment to support wetland habitats and turbid conditions
- riparian vegetation for bank stability, channel structure, food web support, and shade
- floods to maintain seasonal connectivity to upstream habitats
- floodplains that serve as spawning, rearing, and refuge habitat

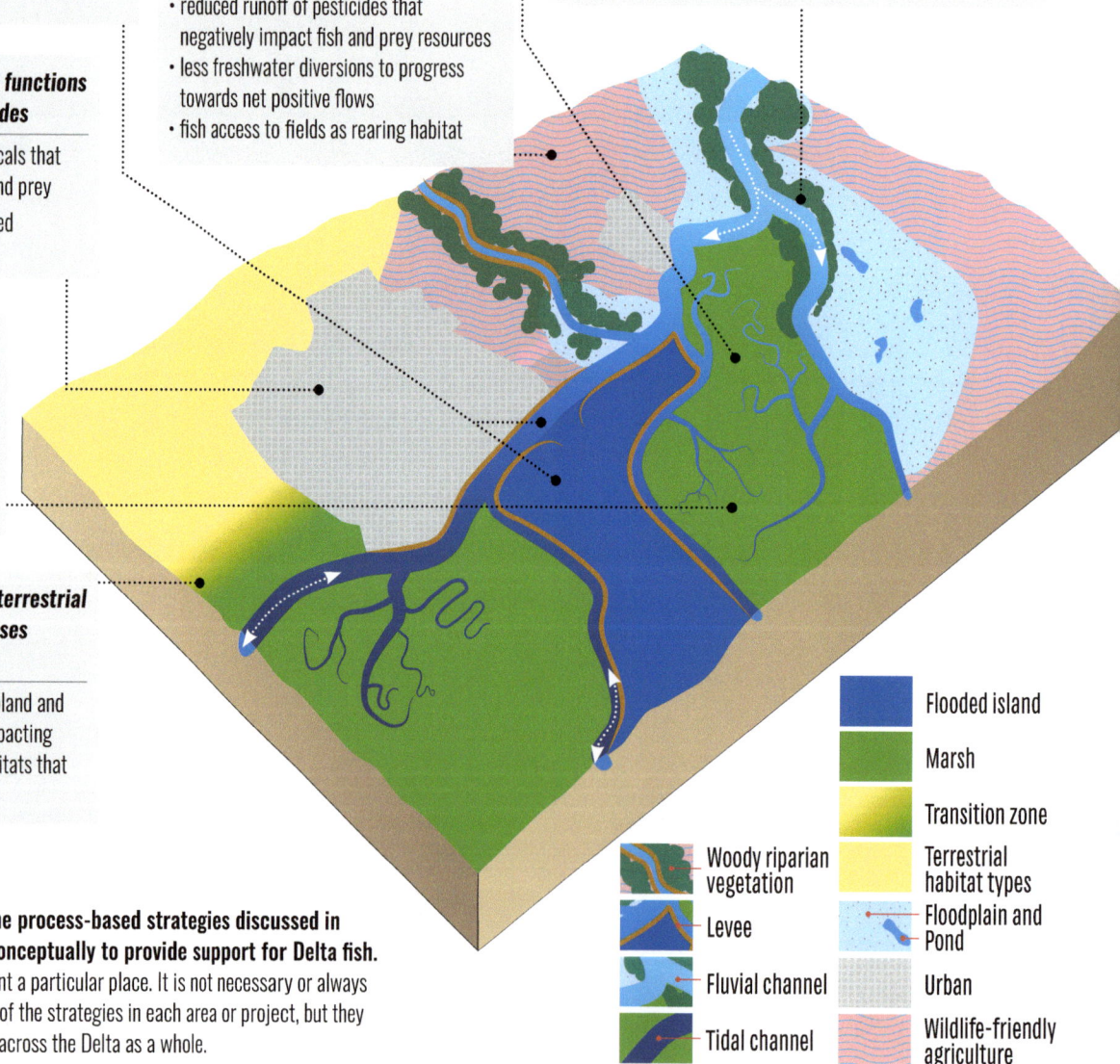

Legend:
- Woody riparian vegetation
- Levee
- Fluvial channel
- Tidal channel
- Flooded island
- Marsh
- Transition zone
- Terrestrial habitat types
- Floodplain and Pond
- Urban
- Wildlife-friendly agriculture

This figure shows how the process-based strategies discussed in Chapter 4 fit together conceptually to provide support for Delta fish. The figure does not represent a particular place. It is not necessary or always appropriate to integrate all of the strategies in each area or project, but they should be well represented across the Delta as a whole.

SFEI AQUATIC SCIENCE CENTER

Potential landscape configuration to support native fish

Re-creating a complex and variable aquatic landscape that supports native fish will require sufficient naturalistic flows and wetland (including floodplain) restoration in many parts of the Delta. The northwest and western sections of the Delta currently offer the greatest potential to support native fish due to the comparatively high flows, and the potential for large marshes and floodplains relative to other parts of the Delta. To better support native fish in these areas, the amount of tidal marsh and fluvial floodplain should be increased, as well as the duration of flooding on floodplains. Increasing riparian cover along the Sacramento River and its tributaries, particularly along Elk Slough, where existing riparian cover is relatively continuous yet narrow, will improve fish support. The Cosumnes River also provides relatively extensive woody riparian forest and scrub, marsh, and floodplain habitat types, which could be increased in the future to improve fish support. While the South Delta currently supports few native fish, the restoration of marshes and other floodplains could provide seasonal benefits to fish in the near term, and might provide year-round support for native fish in the future if exports were decreased to limit reverse flows and reduce entrainment risk. Restoring marshes and complex channel networks, and practicing fish-friendly agriculture wherever possible in the Central Delta, will increase connectivity between areas with more wetlands and floodplains.

Major uncertainties and knowledge gaps

- **Sediment supply:** How much sediment is needed to maintain turbidity conditions beneficial for native fish and to maintain marshes and floodplains as sea level rises? How will this necessary sediment supply be maintained in the long run?

- **Effect of large-scale marsh restoration on turbidity:** Will marshes increase or decrease turbidity? Will the magnitude of change affect native species?

- **Effects of invasive SAV/FAV:** Is SAV/FAV impeding export of marsh production? How can invasive SAV/FAV be controlled or managed to favor support of native aquatic species?

- **Fish use of the marsh:** To what extent do different fish species rely on marsh production? What scale of marsh restoration is needed to improve native fish population viability? Do fish in small sloughs provide a trophic relay of resources to larger sloughs?

- **Water temperatures:** How will fish be affected by increases in water temperature associated with climate change? How effective are suggested measures for mitigating changes in water temperature?

- **Flows and flooding to support native fish:** Will managing lakes as intermittent wetlands favor native species? Will flooded islands provide productive habitat for native fishes if managed properly? How does wastewater from treatment plants affect fish?

- **Limiting factors for support of native fish** including non-listed species such as hitch, Sacramento blackfish, tule perch. Would diked and gated rearing ponds expand populations? Can these populations thrive in less managed habitat types, or will they require intensive management?

(see Appendix C for a more detailed list of uncertainties and knowledge gaps)

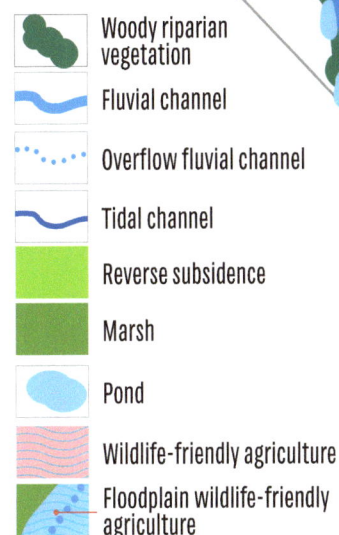

Legend

- Woody riparian vegetation
- Fluvial channel
- Overflow fluvial channel
- Tidal channel
- Reverse subsidence
- Marsh
- Pond
- Wildlife-friendly agriculture
- Floodplain wildlife-friendly agriculture

This conceptual map shows a hypothetical configuration to illustrate how some of the strategies and recommendations might play out at the full Delta scale to support resilient native fish populations.

Supporting **marsh wildlife** in the future Delta

Historically, the vast tidal and non-tidal freshwater emergent wetlands (marshes) of the Delta supported abundant wildlife, including marsh birds and mammals (e.g., rails, herons, bitterns, songbirds, mice, shrews, voles, river otters, and beavers), herpetofauna (e.g., giant garter snakes and chorus frogs), and many species of waterbirds and fish (see separate sections on fish and waterbirds).[12] Many marsh species have experienced tremendous declines in the Delta, some of which are listed as special-status species today (e.g., California Black Rail, California Least Tern, giant garter snake).[13] These declines resulted largely from the loss of 98% of the historical marsh. The smaller, more isolated, and less diverse marshes that remain support smaller wildlife populations, provide fewer resources, and limit the niches and microhabitats available to marsh species.[14] Restoring large populations and robust communities of marsh wildlife in the future Delta will require the re-establishment of large areas of complete, connected marsh. Re-establishing even a relatively modest fraction of the historical marsh extent would provide tremendous benefit to many species.

Successfully re-establishing new marshes and maintaining existing marshes in the context of climate change, water demand, and increasing population growth and development pressure will require us to plan and manage these systems with ecological resilience in mind. This means ensuring marsh persistence in the face of sea-level rise by supporting the accretion and migration processes that allow marshes to increase in elevation.[15] It also means fostering the hydrological and geomorphic processes, and the biological feedbacks, that shape topography and vegetative structure within marshes.[16] Maintaining diverse and connected marsh habitats will provide options for wildlife as conditions change. Creating multiple large marsh areas in the future Delta is critical for achieving sufficient wildlife population sizes, habitat complexity that supports genetic and phenotypic diversity, and source populations for recolonization in the event of local extinction. Less isolated marshes will allow gene flow and wildlife movement.[17] Diversity across the Delta supports adaptation, and is likely to increase resilience through the "portfolio effect," whereby aggregate systems are often less volatile than their components.[18]

Novel habitat types, particularly managed wetlands and wildlife-friendly agriculture, provide additional resources for marsh wildlife. Managed wetlands, as we define them in this report, are designed to provide a limited set of ecological support functions (e.g., seaonal wetlands managed for waterfowl, tule farming for carbon sequestration). These wetlands provide critical support for some marsh species (e.g., waterfowl),[19] but may not benefit others. The benefit of these habitats to marsh wildlife could be increased by integrating them into landscape-scale planning. Key co-benefits of marsh restoration, particularly carbon sequestration and flood protection, are important to consider when setting priorities for land use. Landscape-scale planning will help recover ecological function efficiently, and may help realize these multiple benefits with reduced investment and minimized disruption to other land uses.

Cosumnes River Preserve, photograph by Shira Bezalel (SFEI-ASC)

Recommendations

- **Re-establish and protect large areas of marsh.** Marshes should be large enough to support complete channel networks, hydrologic heterogeneity on the marsh surface, complex plant communities, and viable wildlife populations with high genetic diversity. See "tidal zone processes" (pp. 44-55).

- **Restore or emulate natural processes** that support marsh function and persistence, particularly flows and sediment delivery that support channel formation, marsh-plain heterogeneity, and accretion.

- **Restore broad, complex, continuous marsh-terrestrial T-zones.** Such T-zones will support marsh migration; the exchange of materials between terrestrial and aquatic areas; and wildlife refuge, acclimation and adaptation.

- **Create connectivity between marshes** across the Delta. Marshes should be near one another, and habitat types between marshes should provide supplemental resources and landscape permeability to marsh wildlife whenever possible (e.g., support for Tricolored Blackbirds, giant garter snakes, and waterbirds in agricultural fields). Well-connected marsh habitats can better support marsh wildlife movement, including dispersal, gene flow, and moving to new areas as conditions change (e.g., marsh drowning, new species invasions).

- **Maintain habitat complexity and diversity** by creating large marshes across wide areas of the Delta. The different microclimates and hydrologic patterns that exist in different parts of the Delta support marshes with different channel structure, vegetative communities, and habitat adjacencies. The historical South Delta marsh mosaics and North Delta flood basin marsh types are not well-represented today. Tidal marshes should include diverse freshwater plant species, willow-fern complexes, variations in marsh stature, and natural "duck ponds."

- **Restore complex dendritic channel networks** that increase the amount and complexity of connectivity between marsh and open-water systems. The development of complex channel networks requires the restoration of large marsh patches with full tidal action.

- **Maintain unleveed connections between marsh and open water** to support exchange of materials, energy, and biota. This connectivity allows for export of marsh production to adjacent open-water areas, allows fish access to the marsh plain, and supports complex microhabitats and flooding gradients within marshes that influence vegetative structure and community composition.

- **Create redundancy in support for marsh wildlife** by maintaining multiple large habitat areas, with enough isolation between such areas and populations to reduce the likelihood that one event or stressor (e.g., levee failure, disease, fire, invasive species) would lead to catastrophic loss.

- **Restore marshes near existing remnant and restored marshes** that can serve as propagule sources. The small remnant islands in the Central Delta represent some of the oldest marshes in the Delta and may contain unique genetic diversity.

- **Manage invasive species when necessary.** While most non-native species are not known to be harmful, some invasives will require active management to reduce harmful effects.

Short-term to long-term planning

Sea-level rise increases the urgency of restoring tidal marshes soon to take advantage of large areas that are currently at intertidal elevations. Establishing marshes before sea-level rise accelerates after mid-century will increase the likelihood of their long-term persistence. In the future it may be possible to establish marshes in areas where restoration is not feasible now, particularly upslope of existing or planned marshes along the Delta periphery. Land-use and restoration decisions should consider the possibility of future marsh establishment in these areas currently above intertidal elevations. Marsh restoration, levee breaches, and other management actions have the potential to alter tidal range, necessitating analysis of hydrodynamic effects.

How strategies fit together to support marsh wildlife

Restoring the tidal processes that create tidal marshes and the fluvial processes that support non-tidal marshes is vital to supporting marsh wildlife in the future Delta. Creating complete, complex systems will also require restoring appropriate transition-zone and terrestrial processes, often less considered in marsh restoration. Creating a coherent, integrated landscape that supports marsh wildlife will require us to strategically integrate marsh wildlife support into more developed lands, particularly agricultural areas. Integrating wildlife-friendly agriculture into landscape-scale planning could maximize benefits to wildlife.

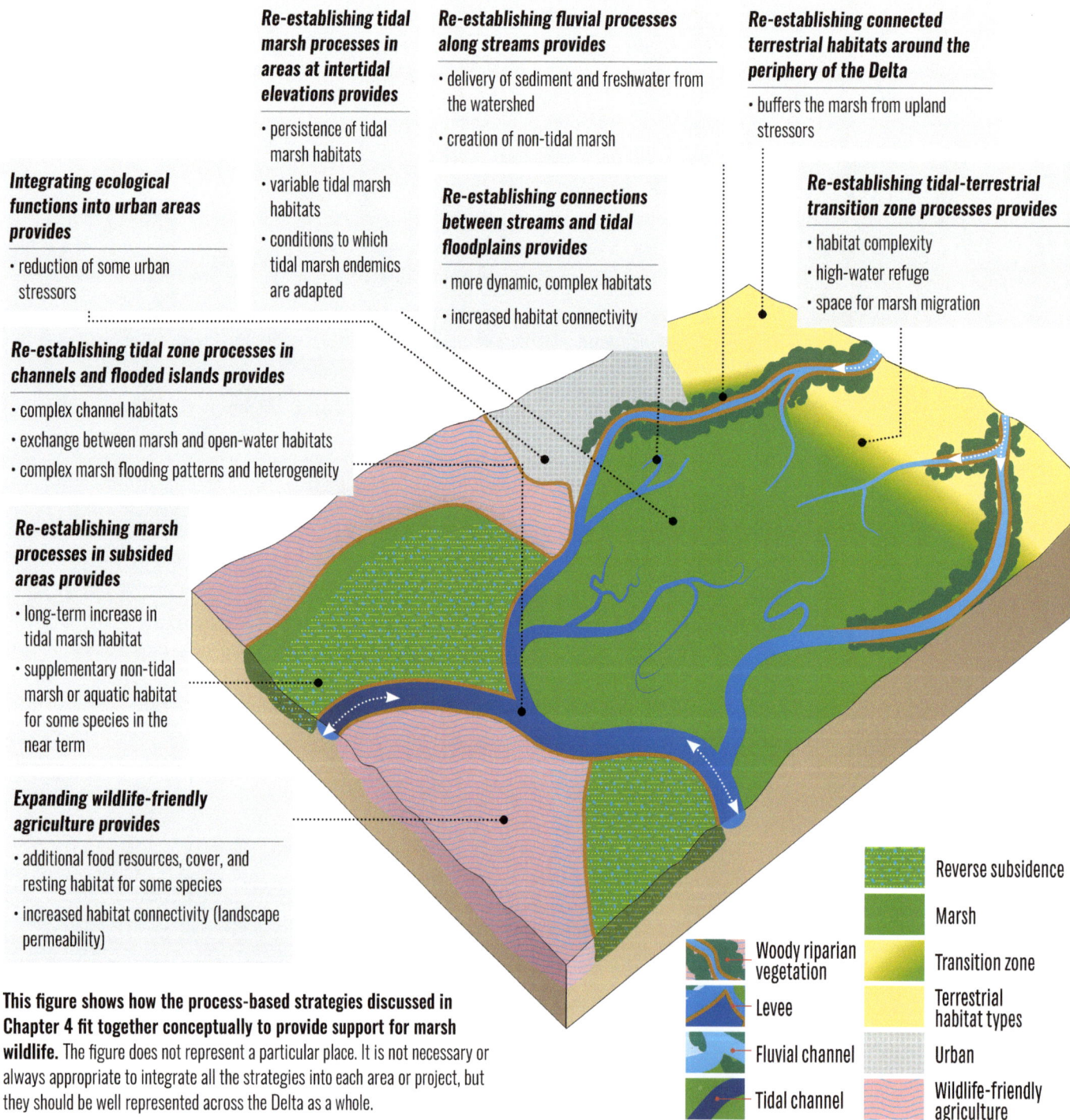

Re-establishing tidal marsh processes in areas at intertidal elevations provides

- persistence of tidal marsh habitats
- variable tidal marsh habitats
- conditions to which tidal marsh endemics are adapted

Re-establishing fluvial processes along streams provides

- delivery of sediment and freshwater from the watershed
- creation of non-tidal marsh

Re-establishing connected terrestrial habitats around the periphery of the Delta

- buffers the marsh from upland stressors

Re-establishing connections between streams and tidal floodplains provides

- more dynamic, complex habitats
- increased habitat connectivity

Re-establishing tidal-terrestrial transition zone processes provides

- habitat complexity
- high-water refuge
- space for marsh migration

Integrating ecological functions into urban areas provides

- reduction of some urban stressors

Re-establishing tidal zone processes in channels and flooded islands provides

- complex channel habitats
- exchange between marsh and open-water habitats
- complex marsh flooding patterns and heterogeneity

Re-establishing marsh processes in subsided areas provides

- long-term increase in tidal marsh habitat
- supplementary non-tidal marsh or aquatic habitat for some species in the near term

Expanding wildlife-friendly agriculture provides

- additional food resources, cover, and resting habitat for some species
- increased habitat connectivity (landscape permeability)

This figure shows how the process-based strategies discussed in Chapter 4 fit together conceptually to provide support for marsh wildlife. The figure does not represent a particular place. It is not necessary or always appropriate to integrate all the strategies into each area or project, but they should be well represented across the Delta as a whole.

Legend:
- Woody riparian vegetation
- Levee
- Fluvial channel
- Tidal channel
- Reverse subsidence
- Marsh
- Transition zone
- Terrestrial habitat types
- Urban
- Wildlife-friendly agriculture

SFEI AQUATIC SCIENCE CENTER

Potential landscape configuration to support marsh wildlife

Elevation is fundamental to determining where resilient marsh habitats can be maintained in the Delta, and therefore where marsh wildlife can best be supported. Elevations that are appropriate for supporting tidal marsh today exist primarily along the periphery of the Delta, with many of the islands in the Central Delta now subsided well below sea level. The largest extant marshes are in the West and Northwest Delta, and the widest expanses of land at intertidal elevation that could be restored to tidal action are in the North and South Delta. Inputs from the Sacramento and San Joaquin rivers could contribute sediment to support marsh accretion and habitat complexity. Additional opportunities exist in the East and Southwest Delta, where there are longer expanses of potentially restorable marsh with adjacent edge habitats to support a broad tidal-terrestrial T-zone. Areas upslope of current intertidal elevations could be managed as non-tidal freshwater marshes, seasonal wetlands, or wildlife-friendly agriculture in the short term, and provide space for marsh migration as sea level rises. Restoring marshes across the Delta should lead to more diverse marsh habitats, with complex mosaics of tidal and non-tidal marshes in the South Delta, flood basins in the North Delta, and brackish marshes in the West Delta. Tule farms, managed seasonal wetlands, flooded agricultural fields, and other novel habitats that provide support to marsh wildlife, will likely provide the greatest benefit when in close proximity to established marshes at intertidal elevations. Large areas of tidal marsh in the future Delta are unlikely to be contiguous, so it is important to maintain landscape elements that increase connectivity between marsh patches, particularly smaller stepping stone marshes and terrestrial habitats that marsh wildlife can disperse across, including wildlife-friendly agriculture.

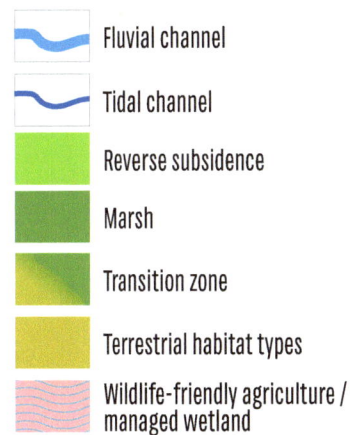

Major uncertainties and knowledge gaps

- **Projections for sea-level rise:** How will tidal range change with sea-level rise? Can we predict in detail how salinity gradients will change?

- **Sediment dynamics:** How much inorganic sediment supply is needed for extant and restored marshes to keep pace with sea-level rise, factoring in peat accumulation? How can sediment deposition in marshes be maximized (or subsidized with sediment from other sources)?

- **Effects of restoration or levee failure on tidal range:** How will opening up large areas of the Delta, particularly in the Central Delta, affect tidal energy in the rest of the Delta? How should restoration be phased or prioritized to balance the urgency of restoration due to sea-level rise with the need to maintain tidal range?

- **Marsh channel re-creation:** How do marsh channels initiate in Delta marsh restoration projects? How do we support formation of dendritic channel networks?

- **Marsh erosion:** How much of a problem is marsh erosion, and where is it happening? What interventions might minimize erosion?

- **Effects of new invasive species:** Which interventions might minimize new invasions?

Legend:

— Fluvial channel

— Tidal channel

Reverse subsidence

Marsh

Transition zone

Terrestrial habitat types

Wildlife-friendly agriculture / managed wetland

This conceptual map shows a hypothetical configuration to illustrate how some of the strategies and recommendations might play out at the full Delta scale to support resilient marsh wildlife populations.

Supporting **riparian wildlife** in the future Delta

Historically, wide woody riparian corridors snaked along the major rivers deep into the Delta marshes, particularly along the Sacramento and San Joaquin, providing habitat for many wildlife species and serving as important corridors to allow terrestrial wildlife access deep into the Delta.[20] These riparian habitats were heterogeneous and dynamic, with variable hydrographs that maintained complex vegetative structure, varied channel morphology, and connections with floodplains.[21] Many riparian species that once thrived in the Delta are now in decline (e.g., riparian woodrat, riparian brush rabbit, tree bats, native bee pollinators) or extirpated from the Delta altogether (e.g., Western Yellow-billed Cuckoo).[22] Riparian wildlife has suffered from not only the loss of habitat extent (60% decline), but also from the change in habitat type configuration and loss of connection to physical processes and appropriate flows. Remaining riparian forests are narrower, less continuous, and scattershot across the Central Delta in places where they did not exist historically.[23]

Restoring support for riparian wildlife in the future will mean re-establishing wide, continuous riparian forests and scrub with connections to off-channel habitats. Making these riparian habitats resilient will require re-establishing appropriate flows and flooding regimes, and insuring adequate sediment and appropriate nutrient inputs from the watershed. Periodic flooding events help maintain diverse and complex riparian forest and scrub vegetation by reseting successional processes.[24] Beneficial flooding of these riparian habitats, along with adjacent large areas of floodplain or flood basin, can provide flood protection for people. Riparian forest patches can also support native bat species that provide pest control benefits and native bee pollinators.[25] Riparian forests have recreation benefits to the public as well (e.g., hiking, birding, kayaking).

Red-winged Blackbird in valley oak, photograph by Kate Roberts (SFEI-ASC)

Recommendations

- *Re-establish flows and flooding that sustain dynamic woody riparian habitats.* This includes restoring a more naturalistic hydrograph, maintaining sufficient flows and sediment to create within-channel heterogeneity, supporting lateral connectivity between channels and complex floodplains, and designing beneficial flooding that supports disturbance regimes and vegetative succession (see "restore fluvial processes" strategy, pp. 60-65).

- *Re-establish and maintain hydrologic connections to the watershed* that provide appropriate inputs of sediment and allow passage for anadromous fish.

- *Restore wide, continuous woody riparian areas* that support movement of terrestrial wildlife through the landscape and provide habitat for a variety of riparian species, including those dependent on wide riparian forests.

- *Restore diversity of riparian habitats* by re-establishing woody riparian corridors along different rivers and creeks. Restore areas of gallery forest, riparian scrub, and willow thicket.

- *Allow room for rivers to meander* to create complex habitats, including oxbows, via levee setbacks and large restoration projects.

- *Increase support for riparian species in agricultural and urban areas* through restoration and buffering of urban creeks, and through best management practices that allow native wildlife to use and pass through agricultural lands. Hedgerows may have particular potential to support connectivity for riparian wildlife. Remnant forest patches can provide refugia for native pollinators that will support agricultural productivity.

- *Connect woody riparian corridors in the Delta to nearby riparian habitats outside of the Delta,* particularly along the Sacramento and San Joaquin rivers.

- *Maintain appropriate groundwater levels to support riparian habitats,* particularly along the Cosumnes River, where the location of wide riparian forest has shifted in response to changes in groundwater levels.

Short-term to long-term planning

Although opportunities along the Sacramento River are limited to nodes and tributaries in the near term, over the long term, opportunities to connect more continuous areas may become available. Long-term planning will need to include designing for larger beneficial flood events, which will require wide riparian forest and large floodplain areas.

How strategies fit together to support riparian wildlife

Critical to supporting the habitat complexity that riparian wildlife depend upon is the re-establishment of fundamental fluvial processes, especially beneficial flooding and sediment transport. Functioning riparian areas require adequate flows, space, and connectivity. Creating a coherent integrated landscape that supports riparian wildlife will require us to strategically integrate riparian wildlife support into developed lands, particularly agricultural areas.

Re-establishing fluvial processes along streams provides

• resilient, persistent woody riparian habitats

• variable riparian habitats maintained by disturbance regimes

• conditions native riparian species are adapted to

Re-establishing connected terrestrial habitats around the periphery of the Delta provides

• increased value of riparian habitat as a corridor through the landscape

• buffer area for reduction of stressors from developed land

Re-establishing connections between streams and tidal floodplains provides

• functional woody riparian corridors

• dynamic riparian habitat

Expanding wildlife-friendly agriculture provides

• supplemental habitat for some edge and riparian species

• habitat connectivity via hedgerows

Integrating ecological functions into urban areas provides

• increased connectivity for riparian wildlife via urban greening and urban stream restoration

• reduction of stressors

Legend:

- Marsh
- Transition zone
- Woody riparian vegetation
- Terrestrial habitat types
- Levee
- Floodplain and Pond
- Fluvial channel
- Urban
- Tidal channel
- Wildlife-friendly agriculture

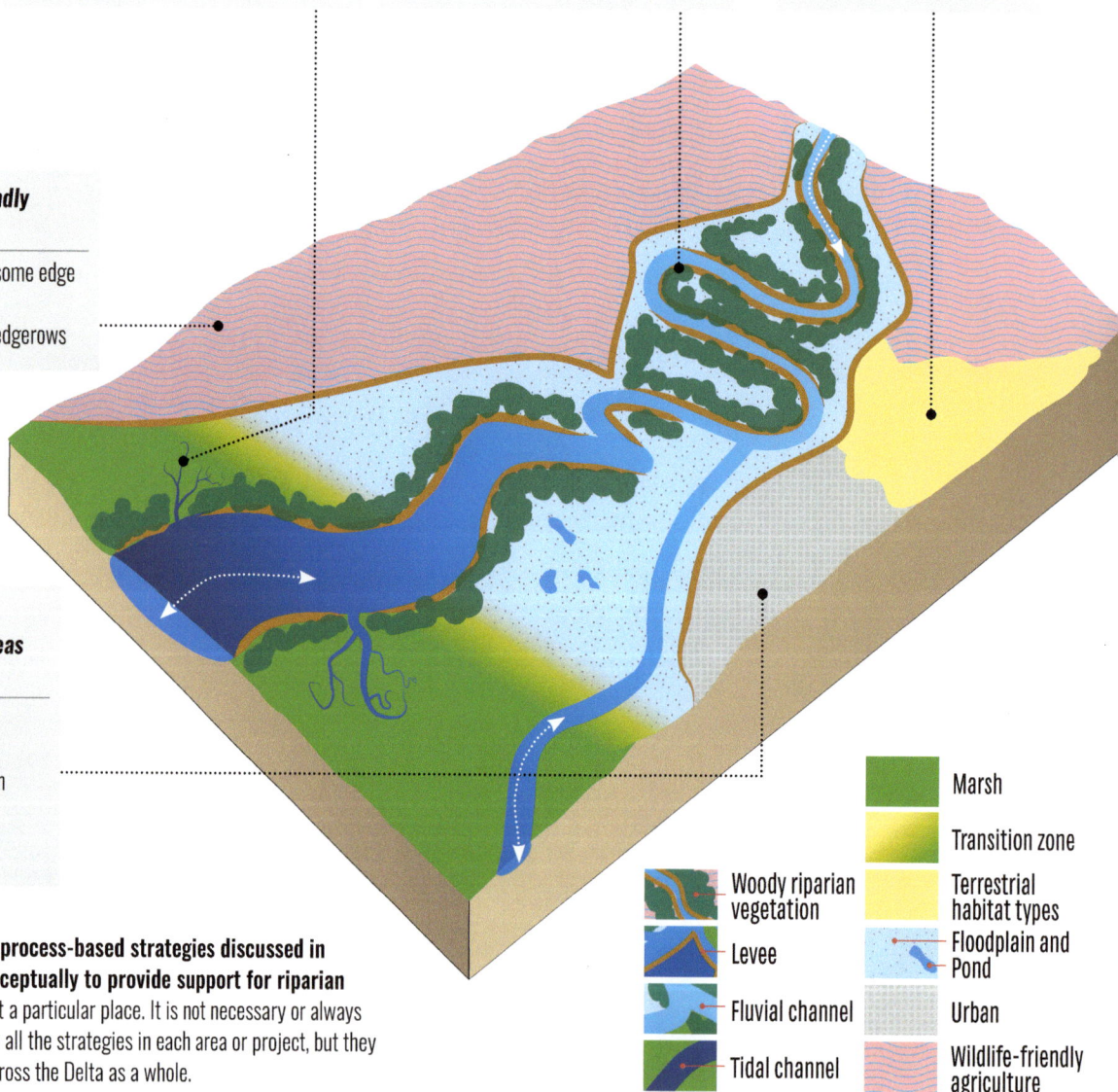

This figure shows how the process-based strategies discussed in Chapter 4 fit together conceptually to provide support for riparian wildlife. It does not represent a particular place. It is not necessary or always appropriate to try to integrate all the strategies in each area or project, but they should be well represented across the Delta as a whole.

Potential landscape configuration to support riparian wildlife

Opportunities to re-establish large, wide areas of continuous woody riparian corridor exist along the Sacramento, San Joaquin, and Cosumnes rivers. The different hydrology, topography, and climate in these areas leads to different types of riparian habitat, with the San Joaquin River supporting more scrub-shrub vegetation, and the Sacramento and Cosumnes rivers supporting gallery forest. Re-establishing willow thickets at the Putah Creek and Cache Creek sinks, and maintaining riparian habitat around Stone Lakes, contributes to the diversity in riparian areas in the Delta. Although opportunities along the mainstem Sacramento River are limited in the short term, opportunities along smaller tributaries exist, particularly along Elk Slough. Remnant topography from historic splays (a deposit of sediment formed when the river broke through it's natural levees) between Elk Slough and the Delta Ship Channel offer opportunities to restore riparian forest in areas adjacent to restored marshes. The currently active floodplains in the Yolo Bypass and along the Cosumnes River could be expanded. There are also floodplain restoration opportunities along the San Joaquin River. Wildlife-friendly agriculture has great potential to increase connectivity in areas adjacent to or in gaps between woody riparian areas.

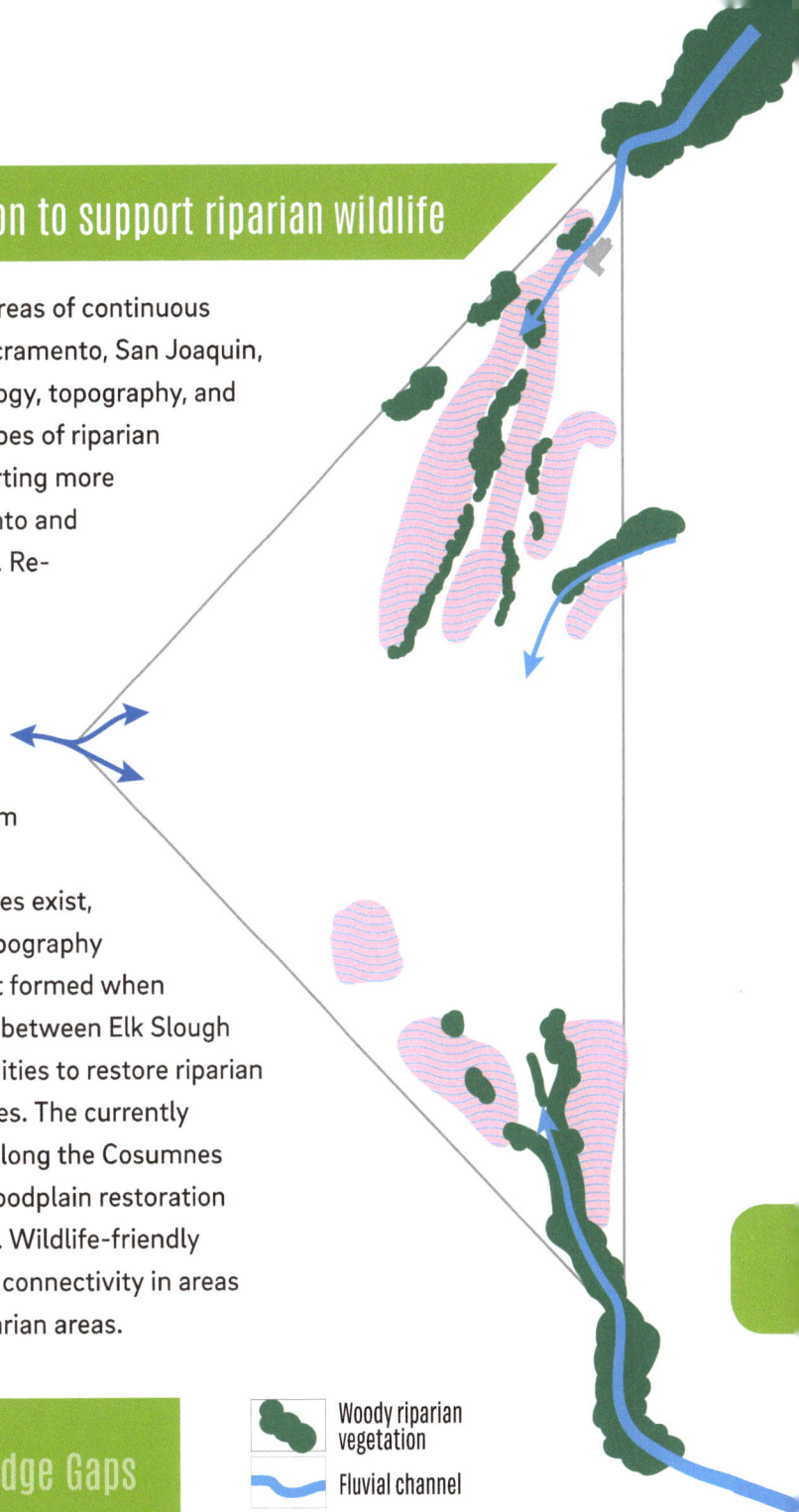

Major Uncertainties and Knowledge Gaps

- **Value of fragmented woody riparian habitats:** What are the relevant gap sizes and riparian corridor widths necessary to support connectivity and habitat for riparian wildlife? What support do small patches of riparian vegetation in the Central Delta provide for wildlife?

- **Temperature regulation:** How much benefit would increased woody riparian shading have on temperature regulation for resident and migratory fish on large and small channels?

- **Alien species:** What is the role of non-native plants and animals in riparian systems?

- **Flows with climate change:** What impact will the change in freshwater flows due to climate change have on riparian systems?

Legend:

- Woody riparian vegetation
- Fluvial channel
- Tidal channel
- Marsh
- Wildlife-friendly agriculture
- Urban

This conceptual map shows a hypothetical configuration to illustrate how some of the strategies and recommendations might play out at the full Delta scale to support resilient woody riparian wildlife populations.

Supporting **waterbirds** in the future Delta

Historically, the Delta supported large numbers of waterbirds, including waterfowl, shorebirds, herons, terns, and grebes.[26] These birds relied on the many different wetland and aquatic habitat types within the Delta. The historical abundance of waterbirds in the Delta was supported by both the diversity and extent of wetlands. For the large number of migrating and overwintering waterbirds that used the Delta, expansive wetlands were critical to providing the necessary food resources to support large populations. Wetlands with short-stature vegetation and shallow standing water, particularly South Delta marshes and seasonal wetlands, probably supported cranes, ibises, and smaller shorebirds. Herons, egrets, and cormorants likely used woody riparian habitats for roosting and nesting. Reduction of the extent and diversity of these habitat types has affected which waterbirds are supported in the Delta.[27]

In the Delta today, waterbirds are critically dependent on agriculture and managed wetlands.[28] Wildlife-friendly agriculture and managed wetlands will remain a key part of the survival of these populations in the future Delta. Increasing the extent of natural wetlands will provide more diverse options for Delta waterbird populations. Waterbirds will require large marshes with abundant food resources, complex channel networks, ponds within marshes, and complex vegetative structure to provide foraging, nesting, brooding, resting, and roosting habitats. Converting managed wetlands that currently support high densities of waterfowl to tidal marsh might reduce the carrying capacity of the Delta for wintering waterfowl. Any such loss might be offset, in part, by increasing the percentage of the agricultural landscape that provides waterfowl benefit. Restoring and protecting wet meadows, woody riparian areas, floodplains, and vernal pools is also critical for supporting many species of waterbirds in the future Delta.

Waterbirds migrating along the Pacific Flyway use wetlands in the Bay, Delta, and Central Valley. Restoration and management actions within this larger region should be coordinated to maximize benefits to waterbirds, and provide life-history support as landscapes evolve with restoration and climate change. Many of the waterbirds in the Delta are charismatic and economically important, supporting hunting, bird watching, and tourism, and inspiring such events as the Sandhill Crane Festival in Lodi and Duck Days in Davis.

Snowy Egret, photograph by Kate Roberts (SFEI-ASC)

Recommendations

- *Restore and maintain a diversity of wetland and aquatic habitat types* including marshes, riparian forests, seasonal wetlands, lakes, and floodplains. Some waterbirds (e.g., egrets, Mallards) can take advantage of many different wetland types, so landscape diversity increases options available to them. For waterbirds with more specialized habitat needs, particular wetland types are critical (e.g., riparian forests for Wood Ducks, short-stature wetlands for Sandhill Cranes, shallow water for shorebirds).

- *Maintain habitat complexity and heterogeneity,* including open water areas within marshes, which are particularly important for wintering and breeding waterfowl. These open water habitat types include channels of various sizes and ponded areas, including those created by beavers. Complex and heterogeneous vegetative structure within wetlands supports roosting and nesting.

- *Restore natural wetlands with short-stature vegetation,* including wet meadows and complex emergent wetlands typical of the South Delta historically. These habitat types likely supported cranes and shorebirds in the past, but few of these wetlands (that are truly analogous to historical conditions) still exist, making it difficult to assess their importance to waterbirds and other wildlife.

- *Restore wetlands of large size* to support adequate food production for large flocks of waterbirds.

- *Coordinate management across the Delta, Bay, and Central Valley* to ensure that there are adequate resources within the larger region as habitats change over time with restoration activities and sea-level rise.

- *Continue to invest in wildlife-friendly agriculture, multi-use floodplains, and managed wetlands,* particularly in areas where the elevation and hydrology support long-term use of these areas. Integrate wildlife-friendly agriculture in to landscape planning, leveraging the support provided by natural wetlands nearby.

Short-term to long-term planning

Wildlife-friendly agriculture, multi-use floodplains, and managed wetlands are critical to support waterbirds in the near term, and will remain important into the future. However, natural wetlands should become an increasingly large portion of the portfolio of habitat types available to waterbirds. Sea-level rise, changing water management, and economic considerations make the long-term future of intensively managed systems unknown. Restoring natural wetlands of the scale and complexity necessary to support waterbirds as conditions change will likely take decades.

How strategies fit together to support waterbirds

Creating the complex, connected habitats necessary to support a diverse suite of waterbirds will require restoring the appropriate tidal, fluvial, and terrestrial-transition zone processes to maintain a diverse mosaic of complementary habitat types. Wildlife-friendly agriculture currently plays a large role in supporting waterbirds in the Delta, and integrating this support into landscape-scale planning would maximize its benefit.

Re-establishing fluvial processes along streams provides

- riparian vegetation for roosting egrets and herons
- riparian forest nest sites for Wood Ducks
- floodplain habitat for shorebirds, waterfowl, other waterbirds
- freshwater marshes as wintering waterfowl habitat

Re-establishing tidal marsh processes in areas at intertidal elevations provides

- wintering waterbird habitat
- waterbird brooding and nesting habitat near ponds and channels

Re-establishing connected terrestrial habitats around the periphery of the Delta provides

- vernal pool habitat for shorebirds
- seasonal wetlands for cranes, waterfowl, shorebirds, herons, and egrets

Re-establishing tidal zone processes in channels and flooded islands provides

- deep-water and benthic invertebrates for diving ducks and other waterbirds
- refuge and brooding habitat in smaller tidal channels

Expanding wildlife-friendly agriculture provides

- habitat for many waterbirds, including critical food resources for wintering waterfowl, roosting habitat for Sandhill Cranes, nesting habitat for shorebirds and terns

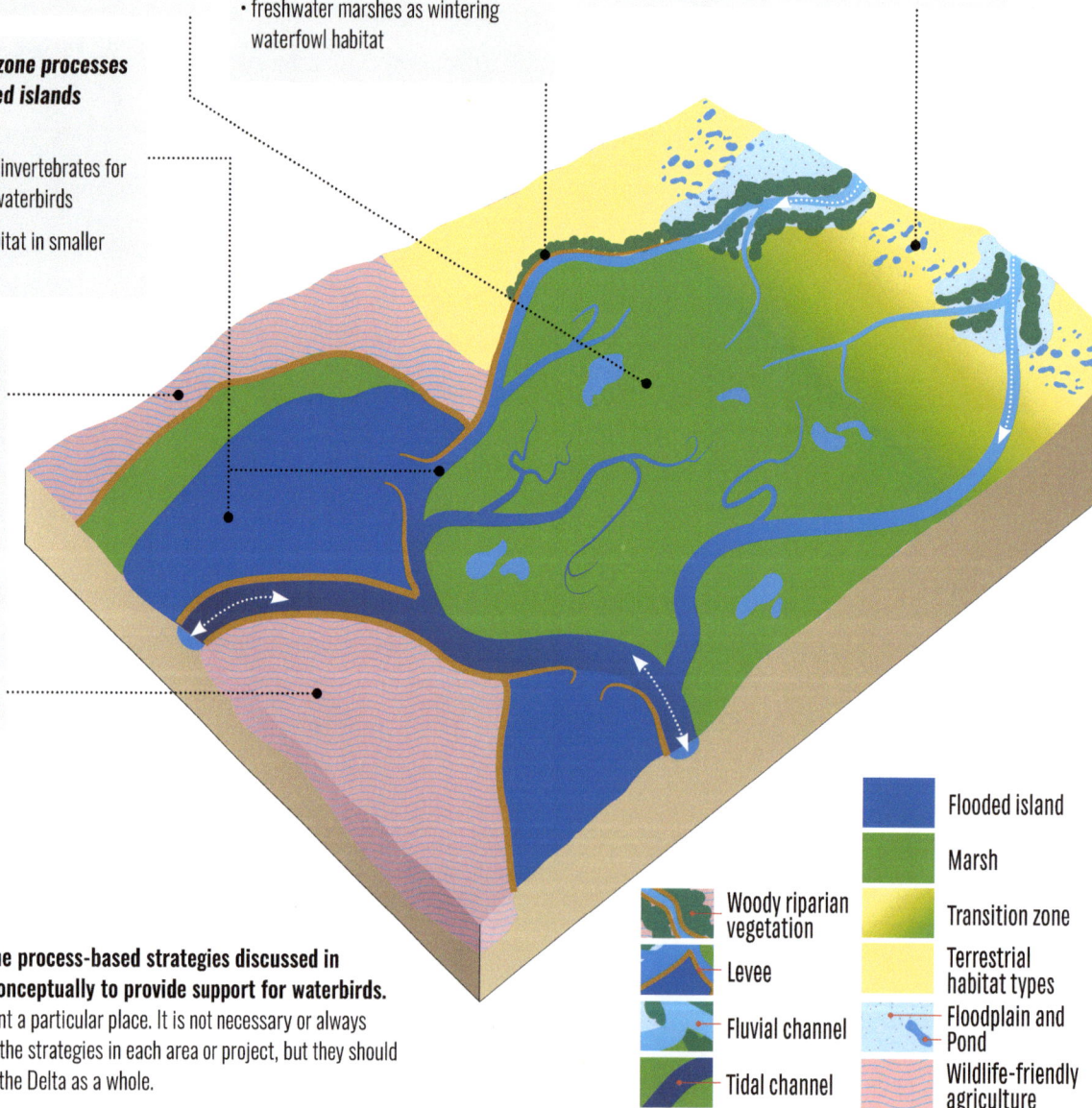

Flooded island

Marsh

Transition zone

Terrestrial habitat types

Woody riparian vegetation

Levee

Fluvial channel

Tidal channel

Floodplain and Pond

Wildlife-friendly agriculture

This figure shows how the process-based strategies discussed in Chapter 4 fit together conceptually to provide support for waterbirds. The figure does not represent a particular place. It is not necessary or always appropriate to integrate all the strategies in each area or project, but they should be well represented across the Delta as a whole.

SFEI AQUATIC SCIENCE CENTER

Supporting numerous and diverse waterbird populations in the future Delta will require an integrated landscape that joins large areas of wildlife-friendly agriculture with a diversity of natural and managed wetlands. Opportunities to create large areas of marsh with complex channel networks exist along the periphery of the Delta, with opportunities to support wide marsh areas in the Cache Slough Complex, Cosumnes River area, South Delta, and West Delta. These large marshes should be of sufficient size to provide food resources for large flocks of overwintering waterfowl. Habitat complexity within these marshes should be geared toward providing roosting, nesting, brooding, and foraging habitat for a variety of waterbirds, including waterfowl, grebes, rails, and terns. Floodplains in the Yolo Bypass and along the Cosumnes and San Joaquin rivers would support shorebirds and dabbling ducks. Riparian and riverine habitats on the Sacramento, San Joaquin, and Cosumnes rivers, as well as on smaller tributaries, would support Wood Ducks, mergansers, herons, and egrets. The Stone Lakes are important for supporting large numbers of waterfowl, and nearby vernal pools and seasonal wetlands are important for cranes and shorebirds. Wildlife-friendly agriculture throughout the Delta can benefit waterbirds, depending on crop types and flooding patterns. The areas along the periphery of the Delta are more likely to be sustainable for waterbird support in the long term as sea level rises.

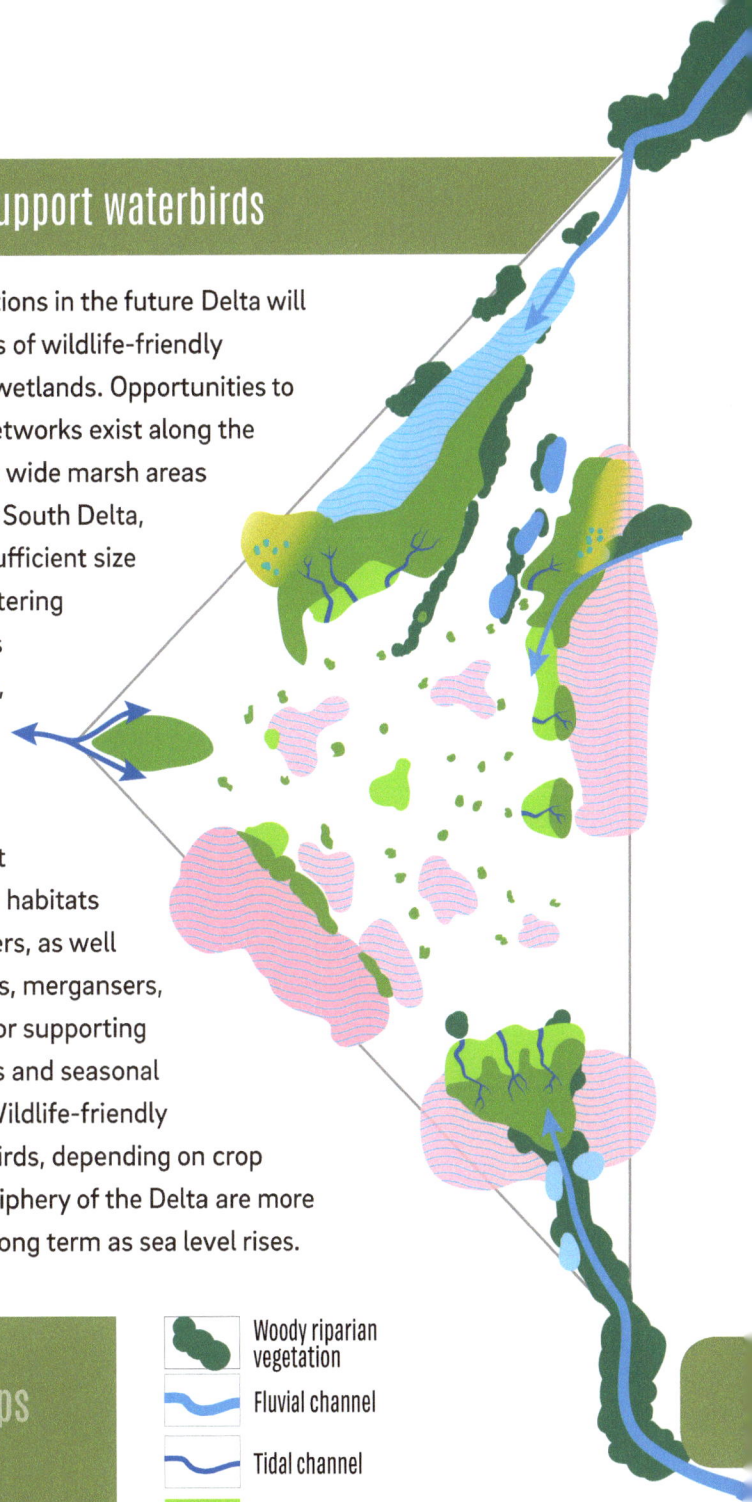

Major uncertainties and knowledge gaps

- **Seasonal wetlands:** Where are conditions appropriate to support large and heterogeneous seasonal wetlands similar to historical habitat types? How much support would these areas provide to shorebirds and other waterbirds?

- **Primary production:** How many overwintering waterfowl can large tidal marshes support?

- **Long-term support:** How can we track and coordinate habitat evolution across the Delta, Bay, and Central Valley over time to ensure long-term support for waterbirds as these regions change?

- **Tradeoffs:** How will conversion of agricultural fields to tidal marsh impact water bird populations?

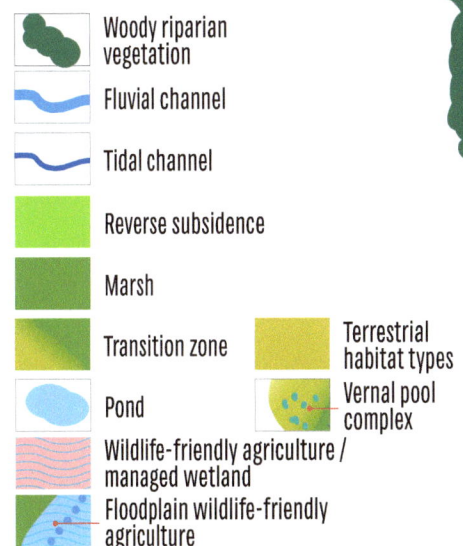

Legend:

- Woody riparian vegetation
- Fluvial channel
- Tidal channel
- Reverse subsidence
- Marsh
- Transition zone
- Pond
- Terrestrial habitat types
- Vernal pool complex
- Wildlife-friendly agriculture / managed wetland
- Floodplain wildlife-friendly agriculture

This conceptual map shows a hypothetical configuration to illustrate how some of the strategies and recommendations might play out at the full Delta scale to support resilient waterbird populations.

Historically, the periphery of the Delta supported a diverse array of wildlife, including species dependent on the estuarine-terrestrial transition zone, species that use both estuarine and terrestrial habitat types, and species associated with terrestrial habitat types around the edge of the historical marsh (e.g., tule elk, bats, giant garter snake, California tiger salamander, Western Burrowing Owl, and California red-legged frog). These terrestrial habitat types included vernal pool complexes, alkali wetlands, seasonal wetlands, grasslands, oak savannas, sand dunes, and riparian forests and scrub.[30] Diverse edge habitat types were associated with unique wildlife communities that contributed substantially to the overall biodiversity within the Delta. Support for edge species in the modern Delta has declined as a result of the loss of wetland and terrestrial habitat types, the fragmented distribution of the remaining edge habitat types, and the replacement of complex, dynamic ecotones with narrow, fixed levees.[31] The leveeing off of islands in the Central Delta has led to a substantial portion of the terrestrial habitats in the modern Delta existing at subtidal elevations, behind levees, where their long-term persistence is in jeopardy.

Supporting resilient terrestrial and transition-zone wildlife communities in the future Delta will mean supporting long, broad, and complex estuarine-terrestrial transition zones and large areas of diverse and connected terrestrial habitat types. Restoration should be prioritized in areas where elevations are appropriate, primarily along the periphery of the Delta, and along the natural levees associated with the Sacramento and San Joaquin rivers, where historical marsh-riparian connectivity could be re-established. Stabilized interior dunes, vernal pool complexes, and alkali wetlands all support many endemic species, have experienced dramatic declines (99%, 73%, and 97%, respectively),[32] and are rare habitat types within the state. The specific soil and hydrology requirements to support these habitat types (e.g., appropriate groundwater levels, hardpans, high soil salinities, sandy soils) make it critical to protect these habitat types where they exist today and preserve or restore the conditions that allow these habitats to persist. Wildlife-friendly agriculture currently provides habitat for many edge species, particularly waterbirds, and this support could be increased in the future Delta through more widespread best management practices and better integration of wildlife-friendly agriculture into landscape-scale planning.

Managing for species in habitat types around the periphery of the Delta is complicated by the cultural and economic importance of these areas to people. Cities and towns in the Delta are primarily located along the periphery of the Delta, or along the high natural levees of the Sacramento River. While this presents challenges, it also affords opportunities. Potential to support edge species in urban areas within or adjacent to the Delta has not been well explored, but it may be possible to improve support for oak savanna- and grassland-associated species in particular.[33]

Black-tailed jackrabbit, photograph by Shira Bezalel (SFEI-ASC)

Recommendations

- *Restore and protect broad, complex, continuous estuarine-terrestrial transition zones.* This requires adjacent, fully tidal marshes to maintain a dynamic backshore and unleveed connections with the watershed.

- *Restore and conserve large areas with diverse terrestrial and wetland habitat types* to support edge functions and a wide suite of species with viable population sizes.

- *Prioritize conserving and restoring terrestrial habitat types along the periphery of the Delta* where elevations are appropriate to maintain habitats above tidal flooding over time.

- *Support landscape configurations that allow high levels of connectivity between areas of the same habitat type* to support dispersal and gene flow among rare endemics (e.g., vernal pool invertebrates, alkali wetland plants).

- *Restore appropriate terrestrial habitat types* for the existing and projected soil, hydrology, and climate conditions. Re-establish the hydrology (including groundwater levels) and soil processes (including hardpan formation) to support these habitat types.

- *Restore sand dunes and riparian habitats adjacent to marsh* to increase the diversity of the estuarine-terrestrial transition zone in the Delta interior.

- *Consider connections between terrestrial habitat types and areas outside the Delta study area.* Important woody riparian, vernal pool, and oak woodland habitats, for example, exist in other parts of the watershed, and are part of the landscape used by Delta wildlife.

- *Increase support for edge wildlife in agricultural lands,* particularly seasonal wetland, grassland, and oak savanna species.

- *Restore more naturalistic seasonal wetlands.* Doing so might require more research, as seasonal wetlands in the Delta today are not necessarily analogous to historical seasonal wetlands, and may support different species.

- *Increase support for sand dune species.* Explore the possibility of restoring former sand dunes where appropriate topography remains. Consider planting dune species in relevant urban areas to increase connectivity and provide habitat for rare plants and invertebrates.

- *Maintain groundwater levels* to support alkali wetlands, seasonal wetlands, vernal pools and other groundwater-dependent systems.

Short-term to long-term planning

In the short term, many areas of seasonal wetlands, grasslands, and other terrestrial habitat types located behind levees on subsided islands will continue to provide critical support for edge species. Over the long term, however, these areas will require increasing investments to maintain levees and drain soils. Resources to support edge wildlife should increasingly be moved toward the periphery of the Delta, to areas where the topography and landscape position are appropriate for long-term habitat resilience.

How strategies fit together to support edge wildlife

Estuarine-terrestrial transition zone and terrestrial processes are essential to maintaining the habitat types that existed around the periphery of the Delta historically, as well as the diverse suite of species associated with those habitat types. Restoring the tidal and fluvial processes that maintain the dynamic edge of these habitat types is also critical. Wildlife-friendly agriculture currently provides support to estuarine-terrestrial edge species, and integrating it into landscape-scale planning could maximize the benefits. Because urban development in the Delta primarily occurs along the terrestrial edge, targeted actions taken in developed areas could help support edge wildlife.

Re-establishing tidal marsh processes in areas at intertidal elevations provides

- dynamic marsh edge to support a broad and productive transition zone
- groundwater recharge

Expanding wildlife-friendly agriculture provides

- supplementary habitat for upland species, particularly oak savanna-associated species
- increased landscape permeability to support wildlife movement through urban and agricultural matrix
- control of urban and agriculture-related stressors

Integrating ecological functions into urban areas provides

- supplementary habitat for upland species, particularly oak savanna-associated species
- increased landscape permeability to support wildlife movement through urban matrix
- control of urban-associated stressors

Re-establishing connected terrestrial habitats around the periphery of the Delta provides

- habitat for edge wildlife, including rare endemics, particularly in peripheral wetlands
- habitat for upland generalists

Re-establishing fluvial processes along streams provides

- woody riparian corridors adjacent to productive marshes
- groundwater recharge

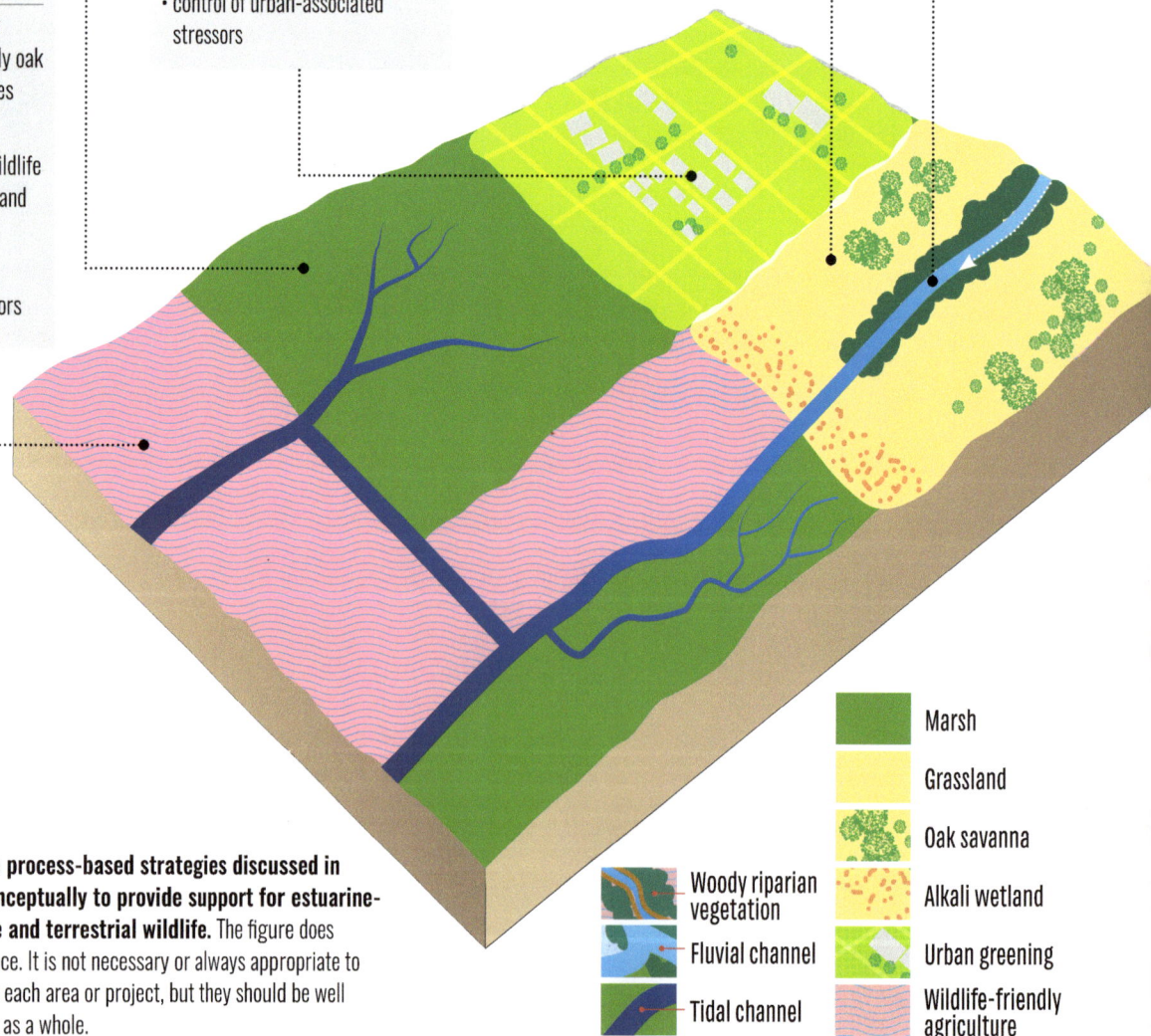

Marsh

Grassland

Oak savanna

Alkali wetland

Urban greening

Wildlife-friendly agriculture

Woody riparian vegetation

Fluvial channel

Tidal channel

This figure shows how the process-based strategies discussed in Chapter 4 fit together conceptually to provide support for estuarine-terrestrial transition zone and terrestrial wildlife. The figure does not represent a particular place. It is not necessary or always appropriate to integrate all the strategies in each area or project, but they should be well represented across the Delta as a whole.

SFEI AQUATIC SCIENCE CENTER

Potential landscape configuration to support edge wildlife

Broad, complex, and continuous estuarine-terrestrial transition zones should be incorporated into future marsh restoration. Opportunities to restore these transition zones exist along the Northwest Delta, where Cache Slough represents the longest stretch of estuarine-terrestrial transition zone. Long stretches of riparian habitat currently line Elk Slough and Prospect Island, and further marsh restoration is planned in the area. Additional opportunities exist in the East and Southwest Delta, where the best opportunities for marsh restoration occur in long strips along the Delta periphery. Additional opportunities exist in the South Delta, where marsh could be restored adjacent to existing woody riparian corridors, and where other edge habitats might be restored as well. Wildlife-friendly agriculture has the potential to support edge species along large swaths of the Delta periphery, particularly for wet meadow and grassland species, but also for oak savanna species where conditions allow. Rare interior stabilized-dune habitat currently occurs within the Antioch Dunes National Wildlife Refuge. Remnant sand dune topography in the West Delta may offer the opportunity to restore sand dunes adjacent to tidal marsh in the future. Vernal pools in the Cache Slough and Stone Lakes areas, and alkali wetlands along the southwest edge of the Delta should continue to be protected, and additional restoration opportunities nearby should be explored. Urban greening and other actions within cities could increase support for edge species, particularly oak savanna and grassland associates.

Major uncertainties and knowledge gaps

- **Seasonal wetlands:** Where can seasonal wetlands that resemble historical habitat types be re-established? What groundwater conditions are needed to support these habitat types?

- **Urban greening to support edge species:** What actions in which areas would provide the most benefit to edge species? How can native oak woodland and grassland species best be supported in urban areas? How can urban greening be implemented to benefit native wildlife and improve other ecosystem services (shading, water quality, improving hydrographs and reducing floods)?

- **Sand dunes:** Can restoration actions in nearby areas support sand-dune species in the West Delta near existing sand-dunes?

- **Alkali wetlands and vernal pools:** How do we expect the conditions and processes that support these habitat types to shift in the coming decades? What potential is there for increased restoration?

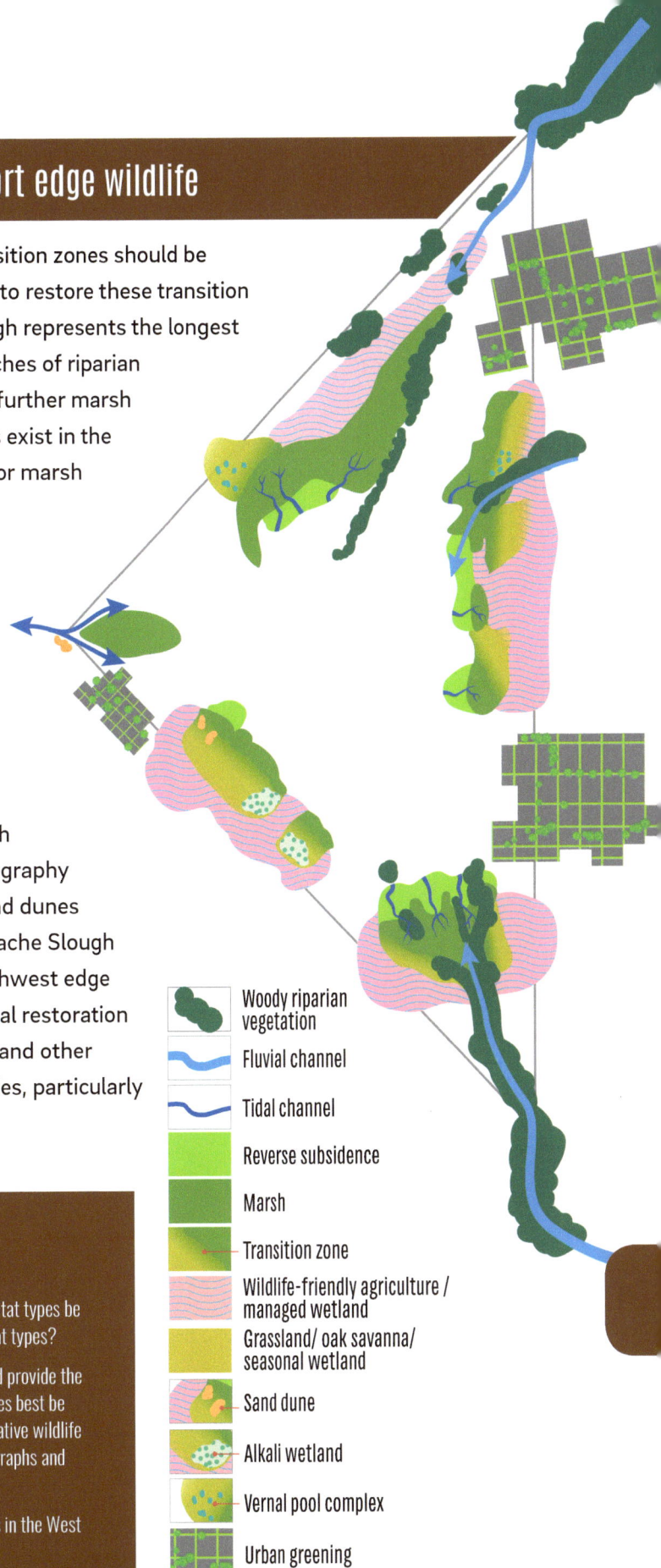

Legend:
- Woody riparian vegetation
- Fluvial channel
- Tidal channel
- Reverse subsidence
- Marsh
- Transition zone
- Wildlife-friendly agriculture / managed wetland
- Grassland/ oak savanna/ seasonal wetland
- Sand dune
- Alkali wetland
- Vernal pool complex
- Urban greening

This conceptual map shows a hypothetical configuration to illustrate how some of the strategies and recommendations might play out at the full Delta scale to support resilient edge wildlife populations.

Primary productivity is a vital ecosystem function that forms the basis of the food web. The potential capacity of ecosystems to support fish, birds, and other wildlife is in large part set by primary production—the supply of food energy and biochemicals required to produce animal biomass. The consequence of low production is to limit the availability of food to consumers.[34] The constraints on primary production, and the relative importance of different primary producer groups, are major ecological uncertainties in the Delta.[35] In the modern Delta, most primary production is contributed by phytoplankton, which produces about 70 grams of new carbon biomass per square meter per year.[36] This ranks Delta phytoplankton production in the lowest 15% of the world's estuaries.[37]

At a workshop in October 2015, scientists met to investigate how the landscape changes outlined in "*A Delta Transformed*" contributed to this low primary production. Two hypotheses that emerged from the workshop were that 1) land-use changes have had a major impact on primary production, and 2) the Delta has been transformed from an ecosystem largely dependent upon marsh-based production (vascular plants and surface algae) to an ecosystem dependent upon production of aquatic plants and algae (see figure below).[38] Freshwater emergent wetlands have been reduced by 98%, while open-water area has increased by 63%.[39] Preliminary calculations for attendant changes in related primary production suggest that annual tidal marsh vascular plant production has decreased by two orders of magnitude (from about 3,800 to 85 kilotons of carbon), while phytoplankton production has doubled (from about 14 to 27 kilotons of carbon).[40] Major assumptions (i.e., around changes in nutrient and sediment loads that might affect rates of production) were made in order to develop these preliminary calculations, and further study is needed to test these hypotheses more rigorously.

Emerging research is shedding light on how changes in Delta channel geometry and hydrology may have impacted primary production.[41] Although the total volume of water in the Delta has increased substantially due to channel widening and dredging, the ratio of autotrophic habitats (within the photic zone)[42] to heterotrophic habitats (below the photic zone) has decreased by nearly half. Approximately 40% of the

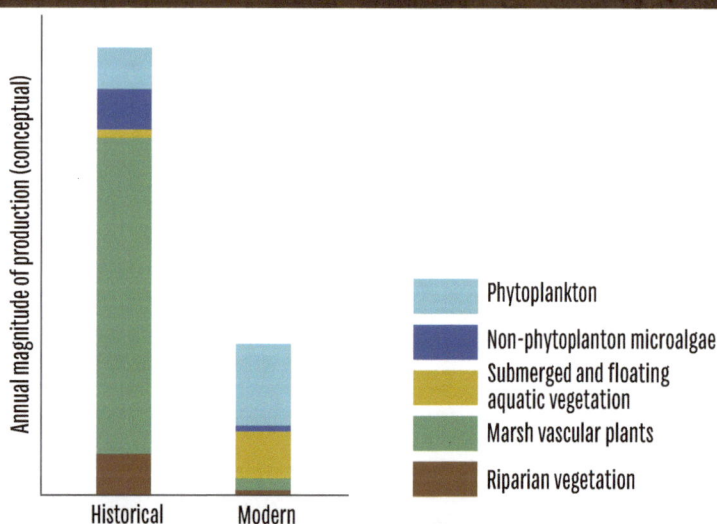

Hypothesized relative contribution of the five primary producer groups to total aquatic primary production based on changes in land use. The size of each bar is proportional to the amount of hypothesized production. Phytoplankton production tracks the increase of open-water area. Marsh vascular plant, non-phytoplankton microalgae, and riparian vegetation production tracks the loss of marsh and riparian habitat types. Submerged and floating aquatic vegetation tracks with evidence that there was limited historical cover of this group, contrasting with large areas of this group today. (Data from Cloern et. al 2016)

Delta's total aquatic habitat volume was autotrophic historically; now only 25% is autotrophic. This change was driven by the loss of shallowly inundated marshes and floodplains, and reduces the relative amount of water-column habitat that can generate food to fuel the food web. The next step in this line of inquiry would be to model phytoplankton production in the context of water moving between autotrophic and heterotrophic habitats and the attendant exchange of nutrients.

Investigating the effects of landscape change on primary production will require 1) rigorous quantitative estimates for a fuller suite of primary producers, 2) information about the nutritional quality of the production, and its transfer efficiency to aquatic consumers, and 3) investigation of the relationship of this production to the hydrodynamics and geometry of the Delta. In addition to the effects of landscape change, major questions remain about the role of nutrients and water quality on productivity in the Delta. Nutrient inputs to the Delta are slated to decrease substantially in coming years due to upgrades in wastewater discharges. The reconstruction of the Sacramento Municipal Wastewater Treatment plant to reduce nitrogen and ammonia is a large-scale experiment in nutrient removal.

A portfolio of food resources

Phytoplankton provides an important food resource in the Delta, but it is not the only important source of primary production for Delta wildlife. Stable-isotope analysis (see right) shows that aquatic consumers in shallow-water habitats eat a variety of food resources, deriving nutrition from many types of primary producers.[43] The fish species sampled eat various foods, likely tracking the shifting availability of food resources throughout the year. Thus, the variety, quantity, and quality of primary production likely all have consequences for Delta consumers. Many birds also rely on aquatic and wetland primary production. Further research is needed to better understand how the portfolio of food resources available to Delta waterbirds has changed over time, and the scale of tidal and non-tidal marsh restoration needed to provide meaningful benefits to wintering waterfowl.

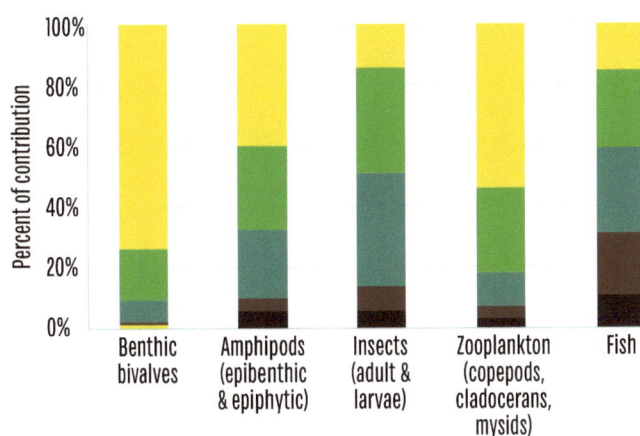

Percent diet contribution of various primary producer groups to primary consumers in shallow-water Delta habitats. Data are from aquatic food webs in vegetated shallow-water habitats. Fish data are from Lindsey Slough. All others are a snapshot from Liberty Island. Data courtesy of Emily Howe.[44]

Legend:
- Benthic diatoms
- Green filamentous algae
- Submerged aquatic vegetation
- Emergent vegetation
- Phytoplankton

Why manage for native biodiversity?

Biodiversity is the variety of life, including variation among genes, species, and ecosystems. It is often measured using richness (the number of unique life forms), evenness (the equitable abundance life forms), and heterogeneity (the dissimilarity among life forms).[45] Maintaining native biodiversity is an important goal of restoration and conservation in the Delta. Recent research on biodiversity draws the link between biodiversity and the maintenance of critical ecosystem processes, such as primary production.[46] Because species perform different ecological roles, complementing each other in a variety of ways, greater diversity can increase the stability of ecosystems, acting as a buffer during periods of drought and other forms of stress.[47]

In supporting the goal of native biodiversity, it is important to manage across the broad suite of ecological functions laid out in the previous pages, creating complex systems with multiple interactions across different taxonomic and functional groups. In addition to the considerations laid out in Chapters 4 and 5, for restoring physical processes and ecological functions, maintaining native biodiversity will also require devoting attention to particular species whose needs require more intensive management. For example, the recommendations made to support native fish (p. 85) are unlikely to sustain the endangered delta smelt without additional measures to address the critically low population sizes and particular stressors increasing mortality.

This approach is in line with "systematic conservation planning" for maximizing biodiversity, which structures conservation targets around three categories: 1) communities and ecosystems (represented in this report by habitat types, physical processes, and ecological functions), 2) abiotic and physical features of special concern, and 3) key species that are likely to be left out by the previous two categories. [48]

Novel ecosystems and non-native species

The San Francisco Estuary, which includes the Delta, is the most invaded aquatic ecosystem in North America.[49] While some invasive species introduced to the Delta have caused major ecological harm (e.g., *Corbicula fluminea*), many non-native species have become a naturalized part of the system, in some cases even providing benefits to native species (e.g., inland silversides that provide critical food resources for native piscivores). In most cases it will not be possible, nor advantageous, to eradicate non-native species from the Delta, so conservation must be reconciled to these new communities. By focusing on native biodiversity, we target recommendations towards bolstering the native component of these novel communities, with the recognition that certain non-native or novel components of the system may provide critical support for native species or additional benefits to people. While naturalized species are an inevitable part of the Delta, managing for existing and future invasive species is a critical part of protecting native biodiversity.

Sandhill Cranes, Cosumnes River Preserve, photograph courtesy Bob Wick (BLM)

06
CONCLUSIONS
& NEXT STEPS

Throughout its history, the Delta has supported both people and wildlife. However, continuing this support in a sustainable way will require a future Delta that looks different from both the present and the past. The Delta cannot remain in its current state; it will inevitably be altered by climate change, levee failure, pressures from competing land uses, and invasive species. Rather than such unintended alterations, we envision a planned future Delta with more space for natural processes and self-sustaining ecosystems, while still remaining a largely agricultural landscape. Restoring the ability of Delta ecosystems to support native wildlife and provide other desired functions cannot be achieved as an afterthought, or through a series of uncoordinated, small-to-moderate sized restoration projects. Rather, restoration and management efforts all need to build toward a shared vision, with larger-scale projects, planning at longer time-scales, and more process-based restoration, conservation and management.

Regional visions are a key next step

This report offers science to inform regional visions that should encompass social and economic considerations, as well as ecological goals, and be stakeholder-driven. Future plans, projects, and management actions can then implement the pieces of these regional visions, synergistically creating a greater cumulative positive impact than just the sum of the parts. Each regional landscape vision should be based in the physical and biological setting of that particular part of the Delta. These visions should encompass short-term and long-term planning, and be adapted over time as lessons are learned and adjustments made. The points below, which are drawn from the Delta Historical Ecology Investigation, A Delta Transformed, and this report, are fundamental to creating effective regional landscape visions for the future Delta.

North Delta, photograph by Kate Roberts (SFEI-ASC)

Artist's rendering of an integrated future for the Delta. In this image, we envision a part of the future Delta where a widened riparian corridor has been integrated into the agricultural landscape, allowing for the development of riparian forests, side channels, and a marshy floodplain along the river. The woody riparian habitats are wide, continuous, and grade down into adjacent marshes, which improves conditions for native riparian wildlife. The marshes are large, with embedded channels that provide access to good rearing habitat for fish during high-water events. These restored habitats would bolster the base of the food web by increasing the abundance and diversity of primary producers, including not just phytoplankton, but also surface algae, marsh vascular plants, and woody riparian vegetation. The habitat patches shown here ideally would connect to other large patches both downstream and upstream, allowing for the movement of wildlife in both directions.

This vision is meant to spark the imagination of the communities who work in and around the Delta. It is not a plan or meant to suggest land-use changes on specific parcels. If we can imagine a more integrated system—building on some of the strategies outlined in this report—we can continue to make progress towards a Delta renewed.

Illustration by Yiping Lu (UC Berkeley)

Different actions are appropriate in different parts of the Delta

The Delta is heterogeneous; some areas are better suited to particular functions than others. The landscapes of the North, Central, and South Delta were created and sustained by different processes, leading to different configurations of habitats and supporting different ecological functions. This heterogeneity extends to the modern Delta as well, though the region looks very different now. Due to the history of subsidence, many of the best opportunities for restoring desired ecological functions now exist along the periphery of the Delta where elevations are appropriate, and tidal, terrestrial, and fluvial processes interact. These areas along the periphery are also highly valued for other human uses, highlighting the need for integrating our ecological goals with social and economic considerations. Around the periphery of the Delta, there are differences in tributary flow regimes, climate, topography/bathymetry, historical habitat types, and wildlife populations that make locally place-based restoration designs critical.

Process-based restoration is the goal, and management will also be required

We focus on process-based restoration, which means restoring or emulating the physical processes (including naturalistic flows, beneficial flooding, and sediment transport) that create and maintain habitats, activate food webs, and support a dynamic landscape. This approach is designed to create habitats that are better suited to the adaptations of native species and are resilient to future change. Given the highly modified Delta landscape and novel species assemblages, process-based restoration will not eliminate the need for ecosystem management. Manipulation of water, sediment, invasive species, and other ecosystem components will be needed in some areas, at varying degrees of intensity. However, process-based restoration should increase the benefits of restoration with less ongoing investment in management, which is important because intensive management is likely to be infeasible at the scale at which restoration must occur.

Actions should support multiple species and ecological functions

We focus on system-level actions designed to benefit a broad suite of native species as the most efficient and effective way to regain desired ecological functions. Targeted actions for rare and endangered species may also be needed to conserve overall biodiversity, and these should be undertaken in the broader context of benefits to native wildlife. Coordinated actions, and consideration of multiple functional groups, can help ensure greater benefits (e.g., projects designed for fish could be adjusted to increase benefits for waterbirds and marsh wildlife as well). Nevertheless, there are likely to be trade-offs and a need for prioritization among goals in specific projects and regions.

North Delta, photograph by Shira Bezalel (SFEI-ASC)

Restoring at large spatial scales is critical for success

Large-scale restoration strategies that incorporate re-establishing flows and floods to Delta rivers and wetlands are more likely to recover the native wildlife support, food-web, and biodiversity functions that have been compromised. Without large areas over which physical and biological processes can occur, achieving viable population sizes and heterogeneous, sustainable landscapes is unlikely. The metrics provided here and in A Delta Transformed can be helpful for planning and evaluating wildlife support at the landscape scale.

Restoration will take time

The Delta has been dramatically altered over the past 200 years. Re-establishing critical physical processes, restoring large areas of habitat, and coordinating myriad separate restoration projects into a coherent landscape will take time, likely decades. Current regulatory, permitting, and funding mechanisms are not designed for such long time scales. Adaptive management, monitoring, and frequent evaluation of progress toward objectives can help us understand and manage interim habitats, as well as adjust trajectories as the landscape evolves and future events unfold.

North Delta, photograph by
Kate Roberts (SFEI-ASC)

Ongoing learning and adjustment are critical

Landscape-scale planning must be an ongoing process that evolves as new lessons are learned and new opportunities arise. Many of the strategies mentioned in this report need to be phased, evaluated, adapted, and honed over time as we become more familiar with how to deploy them. Many research questions still need to be answered (for an initial list see Appendix C: Knowledge Gaps and Information Needs). Coordinated research and monitoring are critical for evaluating restoration success, understanding the changing landscape, and allowing new insights to be quickly applied to future conservation actions. Flexibility and openness to careful experimentation should be built-in to restoration projects. Continued input from stakeholders is essential throughout this process.

Success is possible

Despite the many challenges inherent to achieving large-scale, coordinated restoration in the Delta, there is good cause for hope. Where process-based restoration and actions that integrate wildlife support with agriculture have been undertaken, outcomes have been extremely positive. Restoration and management actions in the Cosumnes River and Yolo Bypass areas cover a relatively small portion of the overall Delta landscape, and they represent a tiny fraction of the historical extent of flooding. However, both areas successfully support critical ecological functions that are desired from the Delta, creating habitat and productive food webs for native wildlife. Such examples suggest that the Delta can be renewed into a place that supports both people and native wildlife sustainably, if we choose to make the investment.

Additional resources are available at the project website:
www.sfei.org/projects/delta-landscapes

APPENDIX A

Defining marshes and wetlands

In this report we use the term "marsh" to refer to habitat types classified as freshwater emergent wetlands in the historical Delta, and their modern equivalents. This includes both tidal and non-tidal freshwater emergent wetlands unless otherwise indicated. This does not include "managed wetlands" that are optimized to provide only a subset of marsh functions – i.e., wetlands managed for waterfowl support or reverse subsidence. "Seasonal wetlands" refers to the wet meadow, vernal pool complex, and alkali seasonal wetland habitat types that occurred along the periphery of the Delta historically, and their modern remnants/equivalents. We discuss wildlife friendly agriculture, including flooded agricultural fields managed temporarily as wetlands, separately, although we recognize that wildlife friendly agriculture can include fields with similar structure and function to either "managed wetlands" or "seasonal wetlands."

Habitat type	Definition
Marsh	Habitat types classified as freshwater emergent wetlands in the historical Delta and their modern equivalents, includes both tidal and non-tidal marshes unless specified
Tidal marsh	Tidal freshwater emergent wetland
Non-tidal marsh	Non-tidal freshwater emergent wetland
Managed wetland	Wetlands optimized to provide only a subset of freshwater emergent wetland functions – i.e., wetlands managed specifically for waterfowl (e.g., duck clubs) support or reverse subsidence (e.g., tule farms)
Seasonal wetland	Wet meadow, vernal pool complex, and alkali seaonal wetland habitat types that occurred along the periphery of the Delta historically and modern remnants/equivalents
Wildlife-friendly agriculture	Practices that support native wildlife on agricultural lands, including practices which manage fields as wetlands that wildlife can access (rice crops and flooded fields)

APPENDIX B:
Species

The table below lists the common and scientific names of the species mentioned in this report. The Integrated Taxonomic Information System (ITIS) database was used as our nomenclatural reference, except for with names marked with a cross (†), which deviate from those validated by ITIS. The common names of all species are written in lower case, with the following exceptions: (1) the common names of all birds are capitalized, as per American Ornithologists' Union standards and (2) all proper nouns are capitalized. Although the word "Delta" is used as a proper noun throughout this report, we do not capitalize the common name of Hypomesus transpacificus (delta smelt), as per U.S. Fish & Wildlife Service standards.

Common name	Scientific name
Birds	
California Black Rail†	*Laterallus jamaicensis coturniculus*
California Least Tern	*Sternula antillarum browni*
Common Yellowthroat	*Geothlypis trichas*
Least Bell's Vireo	*Vireo bellii pusillus*
Mallard	*Anas platyrhynchos*
Marsh Wren	*Cistothorus palustris*
Sandhill Crane	*Grus canadensis*
Snow Goose	*Chen caerulescens*
Song Sparrow	*Melospiza melodia*
Swainson's Hawk	*Buteo swainsoni*
Tree Swallow	*Tachycineta bicolor*
Tricolored Blackbird	*Agelaius tricolor*
Tundra Swan	*Cygnus columbianus*
Virginia Rail	*Rallus limicola*
Western Burrowing Owl†	*Athene cunicularia hypugaea*
Western Yellow-billed Cuckoo†	*Coccyzus americanus occidentalis*†
White-tailed Kite	*Elanus leucurus*
Wood Duck	*Aix sponsa*
Yellow-breasted Chat	*Icteria virens*
Fish	
bluegill	*Lepomis macrochirus*
Chinook salmon	*Oncorhynchus tshawytscha*
delta smelt	*Hypomesus transpacificus*
hitch	*Lavinia exilicauda*
inland silverside	*Menidia beryllina*
largemouth bass	*Micropterus salmoides*
Sacramento blackfish	*Orthodon microlepidotus*
Sacramento splittail	*Pogonichthys macrolepidotus*
Sacramento sucker	*Catostomus occidentalis*
striped bass	*Morone saxatilis*
tule perch	*Hysterocarpus traskii*

Invertebrates

Lange's metalmark butterfly	*Apodemia mormo langei*
overbite clam[†]	*Potamocorbula amurensis*
quagga mussel	*Dreissena bugensis*
zebra mussel	*Dreissena polymorpha*

Mammals

American beaver	*Castor canadensis*
badger	*Taxidea taxus*
California ground squirrel	*Otospermophilus beecheyi*
California vole	*Microtus californicus*
gray fox	*Urocyon cinereoargenteus*
mule deer	*Odocoileus hemionus*
Norway rat	*Rattus norvegicus*
riparian brush rabbit[†]	*Sylvilagus bachmani riparius*[†]
riparian woodrat[†]	*Neotoma fuscipes riparia*
tule elk[†]	*Cervus elaphus nannodes*

Plants

Antioch Dunes evening primrose	*Oenothera deltoides* ssp. *howellii*
Brazilian waterweed	*Egeria densa*
bulrush	*Schoenoplectus* spp.
buckwheat	*Eriogonum* spp.
cattail	*Typha* spp.
Contra Costa wallflower[†]	*Erysimum capitatum* var. *angustatum*[†]
Mason's lilaeopsis	*Lilaeopsis masonii*
milk thistle	*Silybum marianum*
tule[†]	*Schoenoplectus* spp.
water hyacinth	*Eichhornia crassipes*

Reptiles & Amphibians

California red-legged frog	*Rana draytonii*
California tiger salamander	*Ambystoma californiense*
giant garter snake	*Thamnophis gigas*
Western pond turtle	*Actinemys marmorata*

APPENDIX C

Major uncertainties and knowledge gaps related to ecological functions

NATIVE FISH SUPPORT

- *Sediment supply:* How much sediment is needed to maintain turbidity conditions beneficial for native fish and to maintain marsh and floodplain habitats as sea level rises? How will this necessary sediment supply be maintained in the long run?

- *Effect of large scale marsh restoration on turbidity:* Will marshes increase or decrease turbidity? Will the magnitude of change affect native species?

- *Effects of invasive SAV/FAV:* Is SAV/FAV impeding export of marsh production? How can invasive SAV/FAV be controlled or managed to favor support of native aquatic species? Do chemical control measures have negative effects on the primary production of other aquatic organisms?

- *Fish use of the marsh:* To what extent do different fish species rely on marsh production? What scale of marsh restoration is needed to improve native fish population viability? Do fish in small sloughs provide a trophic relay of resources to larger sloughs? Should restoration emphasize bringing native fish to the marshes or marsh production to the fishes?

- *Water temperatures:* How will fish be affected by increases in water temperature associated with climate change? How effective are suggested measures for mitigating changes in water temperature? How large and close together do areas of temperature refuge need to be for significant population-level benefits?

- *Flows and flooding to support native fish:* Will managing lakes as intermittent wetlands favor native species? Will flooded islands provide productive habitat for native fishes if managed properly? How does wastewater from treatment plants affect fish? How will changes to inflows to the Delta with climate change impact fish populations (change in timing, amount of runoff etc)?

- *Limiting factors for support of native fish, including non-listed species such as hitch, Sacramento blackfish, tule perch:* Would diked and gated rearing ponds expand populations? What is the role of the Yolo Bypass in maintaining populations? Can these populations thrive in less managed habitats, or will they require intensive management?

MARSH WILDLIFE SUPPORT

- *Projections for sea-level rise:* How will tidal range change with sea-level rise? Can we predict in detail how salinity gradients will change?

- *Sediment dynamics:* How much inorganic sediment supply is needed for extant and restored marshes to keep pace with sea-level rise, factoring in peat accumulation? How can sediment deposition in marshes be maximized (or subsidized with sediment from other sources)?

- *Effects of tidal marsh restoration or levee failure on tidal range:* How will opening up large areas of the Delta, particularly in the Central Delta, affect tidal energy in the rest of the Delta? How should restoration be phased or prioritized to balance the urgency of restoration due to sea-level rise with the need to maintain tidal range?

- *Marsh channel re-creation:* How do marsh channels initiate in Delta marsh restoration projects? How do we support formation of dendritic channel networks?

- *Marsh erosion:* How much of a problem is marsh erosion, and where is it happening? What interventions might minimize erosion?

- *Effects of new invasive species:* Which interventions might minimize new invasions?

RIPARIAN WILDLIFE

- *Value of fragmented woody riparian habitats:* What are the relevant gap sizes and riparian widths necessary to support connectivity and habitat for riparian wildlife? What support do small patches of riparian vegetation in the Central Delta provide for wildlife?

- *Temperature regulation:* How much benefit would increased woody riparian shading have on temperature regulation for resident and migratory fish on large and small channels?

- *Alien species:* What is the role of non-native plants and animals in riparian systems?

- *Flows with climate change:* What impact will the change in freshwater flows due to climate change have on riparian systems?

WATERBIRDS

- *Seasonal wetlands:* Where are conditions appropriate to support large and heterogeneous seasonal wetlands similar to historical habitats? How much support would these habitats provide to shorebirds and other waterbirds?

- *Primary production:* What density of overwintering waterfowl can large tidal marshes support?

- *Long-term support:* How can we track and coordinate habitat evolution across the Delta, Bay, and Central Valley over time to ensure long-term support for waterbirds as these regions change?

- *Tradeoffs:* How will conversion of agricultural fields to tidal marsh impact water bird populations?

EDGE WILDLIFE

- *Seasonal wetlands:* Where can seasonal wetlands that resemble historical habitats be reestablished? What groundwater conditions are needed to support these habitats?

- *Urban greening to support edge species:* What actions in which areas would provide the most benefit to edge species? How can native oak woodland and grassland species best be supported in urban areas? How can urban greening be implemented to benefit native wildlife and improve other ecosystem services (shading, water quality, improving hydrographs and reducing floods)?

- *Sand dunes:* Can restoration actions in nearby areas support sand-dune species in the West Delta near existing sand-dune habitats?

- *Alkali wetlands and vernal pools:* How do we expect the conditions and processes that support these habitats to shift in the coming decades? What potential is there for increased restoration?

PRIMARY PRODUCTIVITY

Science needs related to effects of landscape change on primary production:
1) rigorous quantitative estimates for a fuller suite of primary producers, 2) information about the nutritional quality of the production, and its transfer efficiency to aquatic consumers, and 3) the relationship of this production to the hydrodynamics and geometry of the Delta.

Nutrients: What were the historical nutrient levels and turbidity conditions? How will future changes in nutrients and suspended sediment/water opacity affect the food web?

BIODIVERSITY

What are the most efficient ways to monitor native biodiversity?

How can the needs of individual, rare, or endangered species be factored into large-scale restoration plans?

ENDNOTES

Chapter 1: INTRODUCTION

1 Moyle et al. 1992.

2 Luoma et al. 2015.

3 Delta Independent Science Board 2016.

4 Lund et al. 2007, U.S. Fish Wildlife Service 2008, California Department of Water Resources et al. 2013, Delta Stewardship Council 2013.

5 CaliforniaWaterFix.com.

6 Prichard 1979.

7 Lund et al. 2010.

8 Delta Independent Science Board 2016.

9 Goals Project 2015.

10 Hellman et al. 2008.

11 CaliforniaWaterFix.com.

12 Cloern et al. 2011.

13 National Resources Council (NRC) 2012.

14 Cloern et al. 2011.

15 National Resources Council (NRC) 2012.

16 Moyle et al. 2012.

17 Whipple et al. 2012.

18 SFEI-ASC 2014.

Chapter 2: GUIDING PRINCIPLES

1 Beller et al. 2015.

2 Goals Project 2015.

3 Delta Stewardship Council 2013.

4 Chelleri and Olazabal 2012, Chelleri et al. 2015.

5 Alberti and Marzluff 2004. Also see Chapter 4, "Integrate Ecological Processes with human land uses."

6 Tompkins and Adger 2004.

7 Simenstad et al. 2006, Hobbs et al. 2014.

8 Anderson and Ferree 2010, Higgs et al. 2014.

9 Stralberg et al. 2011.

10 Chazdon 2003, Cramer et al. 2008.

11 Drever et al. 2006, Beechie et al. 2010.

12 Cloern et al. 2016.

13 Thrush et al. 2008, Standish et al. 2014.

14 Olds et al. 2012, Oliver et al 2013.

15 Gunderson 2000, Fischer et al. 2006, Thrush et al. 2008.

16 Cloern 2007, West and Zedler 2000.

17 Carpenter and Brock 2004, Standish et al. 2014.

18 Hobbs et al. 2006; Moyle et al. 2012.

19 Elmqvist et al. 2003, Standish et al. 2014, Sgro et al. 2011.

20 Ahern 2011.

21 Nyström 2006.

22 Peterson and Holling 1998.

Chapter 3: CONCEPTUAL MODELS OF PHYSICAL AND ECOLOGICAL PROCESSES

1 A list of DREDRIP Delta Conceptual Models is currently maintained at http://www.dfg.ca.gov/erp/cm_list.asp. IEP conceptual models (Sherman et al. [preprint]) are currently in review.

2 Atwater et al. 1979, Malamud-Roam and Ingram 2004.

3 Atwater et al. 1979, Whipple et al. 2012.

4 Atwater et al. 1979.

5 Brown and Pasternack 2005.

6 Stevens and Zelazo 2015.

7 Whipple et al. 2012.

8 Bates and Lund 2013.

9 SFEI-ASC 2014.

10 Luoma et al. 2015.

11 California Department of Water Resources et al. 2013.

11 Smith et al. 2011.

12 Deverel and Leighton 2010.

13 Drexler et al. 2009.

14 Kiernan et al. 2012.

15 Whipple et al. 2012.

16 Williams et al. 2009.

17 Roe and Georges 2007.

18 Semlitsch and Bodie 2003.

19 Roe and Georges 2007.

20 Longcore and Osborne 2015.

21 Elsholz 2010.

Chapter 4: PROCESS-BASED RESTORATION STRATEGIES

1 Wohl et al. 2005, Beechie et al. 2010.

2 Beechie et al. 2010.

3 Booth et al 2016.

4 Kondolf et al. 2006.

5 Wohl et al. 2005.

6 Beechie et al. 2010.

7 SFEI-ASC 2014.

8 San Francisco Estuary Partnership (SFEP) 2015.

9 148,000 ha; Whipple et al. 2012.

10 Our methods for determing areas at intertidal elevation were rudimentary. In absence of a comprehensive spatial dataset indicating the elevations of tidal datums across the Delta, we simply chose a single elevation to represent MLLW (0.64 m NAVD88) and a single elevation to represent MHHW 1.95 m NAVD88) across the study extent. These tidal datum elevations were measured at Cahce Slough by cbec (2010). Eelvations bounding the +3 ft and +6 ft ranges were calculated by adding those amounts to both the current MLLW and MHHW elevations. These methods therefore make the major simplifying assumption that tidal range in the Delta is constant across space and time. We know this assumption is false, and therefore only use this layer to show the approximate location and extent of areas at intertidal elevation now and into the future. The layer should be refined for use in any detailed planning process. A simple visual inspection suggests general agreement with the areas within tidal demarcated by Siegel et al. (2010) using more sophisticated methods that account for spatial variability in the elevation of tidal datums.

11 Whipple et al. 2012.

12 Whipple et al. 2012.

13 Drexler et al. 2009.

14 Drexler 2011.

15 Hood 2007.

16 Beedy 1989 as cited in BRCP 2015.

17 Tsao et al. 2015.

18 N. Nur, personal communication.

19 For each major marsh island in the historical Delta, Whipple et al. 2012 determined the number of major channel networks and total island area. These data were used to calculate, on an island to island basis, the average area per channel network, which ranged from 200 ha of marsh per channel network (Venice Island) to 1,000 ha of marsh per channel network (Sherman Island). The average of these averages (or the average area of marsh per channel network across all Delta islands with major channel networks [n = 9]), was approximately 500 ha.

20 SFEI-ASC 2014.

21 SFEI-ASC 2014.

22 SFEI-ASC 2014.

23 SFEI-ASC 2014.

24 Burau et al. 2007.

25 Johnston 1956.

26 Hall 2015.

27 Kjelson et al. 1981.

28 SFEI-ASC 2014.

29 SFEI-ASC 2014.

30 SFEI-ASC 2014.

31 SFEI-ASC 2014.

32 This expectation is based on the research of Spautz and Nur (2002), who found that, in San Francisco Bay marshes, Black Rails are more likely to be present in patches where core area is large relative to total area. Core area has a significant positive correlation with Black Rail presence, while edge area has a significant negative relationship (Spautz et al. 2005). This said, research has also found marsh patch perimeter to area ratio and fractal dimension (measures similar to core to edge habitat ratio) to not significantly correlate with Black Rail presence (Spautz et al. 2006). The same research found that marsh patch perimeter to area ratio has a significant negative correlation with the presence of other marsh birds (Marsh Wren and Common Yellowthroat).

33 SFEI-ASC 2014.

34 SFEI-ASC 2014.

35 Teal 1962, Odum 1980 as cited in Howe 2014.

36 Eldridge and Cifuentes 2000.

37 Eldridge and Cifuentes 2000.

38 SFEI-ASC 2014.

39 SFEI-ASC 2014.

40 Drexler et al. 2009.

41 Miller et al. 2008.

42 Note that mineral sediment placed on peat soils may sink until the peat below is compacted, a process that must be factored in when considering the direct placement of sediment in subsided areas with remnant peat. Additionally,opportunities for direct placement of sediment are limited by the immense volume of subsided islands and a limited sediment supply. It is estimated, for example, that the construction of the proposed twin tunnels would yield approximately 23 million cubic meters of excavated material (Luery 2013); though this a tremendous volume, it is equivalent to only 0.9% of the Delta's total anthropogenic accommodation space (2.5 billion cubic meters [Mount and Twiss 2005]). 22.9 million cubic meters is, for reference, slightly more than the volume of Frank's Tract (~22 million cubic meters). Restoration activities in the Delta that require sediment or alter the amount/distribution of suspended sediment should be considered in regional sediment-management plans, as recommended in the Baylands Ecosystem Habitat Goals Science Update (Goals Project 2015).

43 Doody 2007 (p. 28) describes "warping."

44 Bates and Lund 2013.

45 The assumed accretion rate of 5 cm/yr is less than the maximum rate of 7-9 cm/yr, but higher than the average rate of 4 cm/yr, measured at the Twitchell Island wetlands (Miller et al. 2008). Subsidence reversal rates have been shown to decrease over time (Anderson et al. 2016), so it is important to note that 5 cm/yr rate might be on the high end of what is achievable.

46 Drexler et al. 2009.

47 Miller et al. 2008.

48 Drexler 2011.

49 c.f., Kirwan and Murray 2007, Schoellhamer et al. 2012.

50 Miller et al. 2008.

51 Wiens 2002.

52 Lund et al. 2007, Enright 2008, Enright and Culberson 2009, Whipple et al. 2012, Safran et al. 2016.

53 c.f., Enright et al. 2013.

54 SFEI-ASC 2014.

55 Campbell Grant et al. 2007, Enright et al. 2013, Morgan-King and Schoellhamer 2013.

56 Lund et al. 2010

57 Moyle et al. 2012.

58 Moyle 2008.

59 T. Sommer and P. Moyle, personal communication.

60 Morgan-King and Schoellhamer 2013.

61 Moyle et al. 2012.

62 Moyle et al. 2012.

63 SFEI-ASC 2014.

64 Enright 2008, Safran et al. 2016, Morgan-King and Schoellhamer 2013.

65 Campbell Grant et al. 2007 and references therein.

66 Whipple et al. 2012, Safran et al. 2016

67 C. Enright, personal communication.

68 Jones et al. 2000 as cited in Campbell Grant et al. 2007.

69 Aquatic Science Center 2012, Luoma et al. 2015.

70 Enright et al. 2013.

71 SFEI-ASC 2014.

72 SFEI-ASC 2014.

73 Enright 2008, Enright et al. 2013, , Morgan-King and Schoellhamer 2013, Safran et al. 2016.

74 SFEI-ASC 2014.

75 SFEI-ASC 2014.

76 SFEI-ASC 2014.

77 See note 73 above.

78 Grant et al. 2007.

79 cf. Morgan-King and Schoellhamer 2013.

80 Bate et al. 2004, Florsheim et al. 2008, Ensign et al. 2013.

81 Florsheim et al. 2008.

82 Barwis 1977, Bate et al. 2004, Florsheim et al. 2008, Ensign et al. 2013, Keevil et al. 2013.

83 Odum 1988, Brinson et al. 1995, Rundle et al.1998, Cavallo et al. 2012. Cavallo et al. note that migration rate of juvenile salmon moving down the Sacramento River decreases when they enter the tidal-fluvial transition zone.

84 Cavallo et al. 2009.

85 Cavallo et al. 2009.

86 Matella and Merenlender 2014, Yarnell et al. 2015.

87 Dettinger et al. 2004.

88 SFEI-ASC 2014.

89 Dusterhoff et al. 2014, Goals Project 2015.

90 Dusterhoff et al. 2014.

91 Goals Project 2015.

92 Elsholz 2010 notes high mortality of riparian brush rabbit during floods in an area with minimal high water refuge.

93 SFEI-ASC 2014.

94 Previously unpublished data from SFEI-ASC 2014.

95 Previously unpublished data from SFEI-ASC 2014.

96 Ingebritsen et al. 2000.

97 Yoshiyama et al. 2001.

98 Whipple et al. 2012.

99 Whipple et al. 2012.

100 Swanson 2015.

101 Dettinger and Cayan 1995.

102 Das et al. 2011.

103 Dettinger et al. 2004.

104 Yarnell et al. 2015.

105 Florsheim et al. 2008.

106 Kondolf and Williams 1999, Kondolf and Wilcock 1996.

107 Richter et al. 2011.

108 State Water Resources Control Board 2010.

109 San Francisco Estuary Partnership (SFEP) 2015.

110 State Water Resources Control Board 2010.

111 Florsheim et al. 2008.

112 Brett 1952

113 State Water Resources Control Board 2010.

114 Matella and Merenlender 2014; note that, while theMatella and merenlender report a minimum value of 21 days for splittail spawning and rearing, a summary of the literature prepared by Swanson et al. (2010) reports that benefits of inundation to splittail abundance do not begin to occur until inundation duration exceeds 30 days, while maximum benefits probably occur when the duration exceeds 45 days, both of which are longer than the minimum value reported by Matella and Merenlender.

115 SFEI-ASC 2014.

116 Sommer et al 1997.

117 Sommer et al. 1997 and Feyrer et al. 2006, as cited in Swanson 2015.

118 Sommer et al. 1997 and Feyrer et al. 2006, as cited in Swanson 2015.

119 Ahearn et al. 2006, Grosholz and Gallo 2006.

120 Boulton et al. 2010.

121 Boulton et al 1998.

122 Whipple et al. 2012.

123 Steiger et al. 2005.

124 Collins et al. 2006.

125 Collins et al. 2006.

126 Laymon and Haltermann 1989.

127 Laymon and Haltermann 1989.

128 San Francisco Estuary Partnership (SFEP) 2015.

129 San Francisco Estuary Partnership (SFEP) 2015.

130 Unpublished data from analyses published in San Francisco Estuary Partnership (SFEP) 2015.

131 Unpublished data from analyses published in San Francisco Estuary Partnership (SFEP) 2015.

132 Zeiner et al. 1990, Gaines 1977 as cited in BRCP 2015.

133 Laymon and Haltermann 1989.

134 Laymon and Haltermann 1989.

135 SFEI-ASC 2014.

136 SFEI-ASC 2014.

137 Cecala et al. 2014.

138 Cecala et al. 2014 and references therein, Christie and Knowles 2015.

139 Cecala et al. 2014.

140 Tremblay and St.Clair 2009.

141 G. Geupel, personal communication as cited in SFEI-ASC 2014.

142 SFEI-ASC 2014.

143 SFEI-ASC 2014.

144 SFEI-ASC 2014.

145 Whipple et al. 2012.

146 Opperman et al. 2010, Whipple et al. 2012.

147 Elsholz 2010, Schneider et al. 2012.

148 Goals Project 2015; note that we modified the T-Zone definition published in the Goals Project Update by removing the estuarine-fluvial transition component, which we instead address in our "Restore tidal-fluvial transition-zone processes" strategy.

149 Whipple et al. 2012.

150 Many of the ecological functions provided by T-Zones in the Delta are described in SFEI-ASC 2014. Another reference is Goals Project (2015; Science Foundation Chapter 4). Though the focus of this latter source is on the Bay, many of the functions and ecosystem services it describes also apply to T-Zones in the Delta.

151 San Francisco Estuary Partnership (SFEP) 2015.

152 Goals Project 2015 (see 'Management of Marsh-Upland Transitional Habitats' by Fulfrost et al.).

153 Kneib et al. 2008 and references therein.

154 Odum 1990.

155 Goals Project 2015.

156 SFEI-ASC 2014.

157 New analysis based on data from SFEI-ASC 2014.

158 SFEI-ASC 2014, Goals Project 2015.

159 New analysis based on data from SFEI-ASC 2014.

160 New analysis based on data from SFEI-ASC 2014.

161 Goals Project 2015.

162 Semlitsch and Bodie 2003.

163 Collins et al. 2007.

164 Collins et al. 2007.

165 Kuerzi 1941 in Collins et al. 2007.

166 Semlitsch and Bodie 2003.

167 Kelly et al. 2008.

168 SFEI-ASC 2014.

169 SFEI-ASC 2014.

170 Hobson and Dahlgren 1998. Also see Rains et al. 2008, which discuss the importance of geology in controlling soil formation processes in different vernal pool habitats.

171 Solomeshch et al. 2007 discuss the importance of groundwater depth and duration on the species composition of vernal pools.

172 Allen-Diaz et al. 2007.

173 Whipple et al. 2012.

174 Rains et al. 2005.

175 Elmore et al. 2006, McLaughlin and Zavaleta 2012.

176 Rains et al. 2005.

177 McLaughlin and Zavaleta 2012.

178 Whipple et al. 2012.

179 Smith and Verrill 1998.

180 The maximum distance Pilliod et al. (2013) found a turtle from a pond on the Carrizo Plain was 345 m. A circle with a radius of 345 m has an area of 37 ha.

181 Searcy and Shaffer (2011) conclude that 2,092 m is a biologically-justified maximum migration distance of California Tiger Salamander away from their breeding pools. A circle with a radius of 2,092 m has an area of 1,372 ha.

182 BRCP 2015.

183 Kie et al. 2002 as cited in BRCP 2015.

184 Dunk and Cooper 1994.

185 California Department of Water Resources et al. 2013.

186 Messick and Hornocker 1981 as cited in BRCP 2015.

187 Gervais et al. 2000 as cited in BRCP 2015.

188 Estep 1989 as cited in BRCP 2015.

189 Semlitsch and Bodie 2003.

190 Searcy and Shaffer 2011.

191 Mawdsley et al. 2009.

192 http://www.delta.ca.gov/Landscapes.htm

193 Elphick et al. 2010, Katz et al. 2013.

194 Ivey et al. 2011.

195 e.g., Herzog 1996, Katz et al. 2013, http://ca.audubon.org/birds-0/tricolored-blackbirds.

196 Boyles et al. 2011

197 Chaplin-Kramer et al. 2011.

198 van Groenigen et al. 2003.

199 http://www.nrcs.usda.gov/wps/portal/nrcs/main/ca/programs/farmbill/

200 Sullivan et al. 2014; Meadows 2014.

201 Katz et al. 2013, Conrad et al. 2016.

202 Hinsley and Bellamy 2000.

203 Grossinger and Whipple 2009, Whipple et al. 2010 (see "Opportunities for Restoration" section).

204 Katz et al. 2013, Conrad et al. 2016.

205 M. Kepner, personal communication.

206 G. Yarris, personal communication.

207 Sabalo 2016.

208 e.g., Weston et al. 2014.

209 Budd et al. 2009, Zhang and Goodhue 2010.

210 Monson et al. 2007.

211 Moyle et al. 1992.

212 Vogel 2013.

213 Delta Protection Commission 2013.

214 http://www.nature.org/ourinitiatives/regions/northamerica/unitedstates/california/howwework/california-migratory-birds.xml

215 SFEI-ASC 2014.

216 Delta Protection Commission 2013, R. Kelsey and J. Beringer, personal communication.

217 Hinsley and Bellamy 2000.

218 Davies and Pullin 2007.

219 Gilroy et al. 2014.

220 Wu 2014 and references therein.

221 Grossinger and Whipple 2009, Whipple et al. 2010 (see "Opportunities for Restoration" section).

222 Dietz 2007.

223 Gersberg et al. 1986, Bastian and Hammer 1993.

224 Aquatic Science Center 2012; 303(d) list.

225 Jokimäki 1999.

226 Abensperg-Traun and Smith 1999.

227 SFEI-ASC 2014.

228 McKinney 2006 and references therein.

229 Tzoulas et al. 2007.

Chapter 5: SUPPORTING ECOLOGICAL FUNCTIONS IN THE FUTURE DELTA

1 SFEI-ASC 2014.

2 Whipple et al. 2012.

3 Moyle 2002, Feyrer and Healey 2003, Nobriga et al. 2005.

4 SFEI-ASC 2014.

5 Toft et al. 2003, Brown and Micniuk 2007.

6 Jassby and Cloern 2000.

7 Moyle 2014.

8 Nobriga and Feyrer 2007.

9 Sommer et al. 2004.

10 Cloern et al. 2011.

11 Swanson et al. 2000.

12 SFEI-ASC 2014.

13 California Department of Water Resources et al. 2013.

14 SFEI-ASC 2014.

15 Morris et al. 2002.

16 Williams and Orr 2002.

17 Amezaga and Green 2002.

18 Schindler et al. 2015

19 Central Valley Joint Venture 2006

20 SFEI-ASC 2014.

21 Whipple et al. 2012.

22 California Department of Water Resources et al. 2013.

23 SFEI-ASC 2014.

24 Beechie et al. 2010.

25 Boyles et al. 2011; Winfree et al 2007

26 Whipple et al. 2012.

27 SFEI-ASC 2014.

28 Central Valley Joint Venture 2006

29 SFEI-ASC 2014.

30 Whipple et al. 2012.

31 SFEI-ASC 2014.

32 SFEI-ASC 2014.

33 Grossinger and Whipple 2009, Whipple et al. 2010 (see "Opportunities for Restoration" section).

34 Cloern et al. 2016.

35 Cloern et al. 2016.

36 Jassby and Cloern 2000, Cloern et al. 2014.

37 Cloern et al. 2014.

38 Cloern et al. 2016.

39   SFEI-ASC 2014.

40   Robinson et al. 2016.

41   Fleenor et al. (preprint).

42   Autotrophic habitats were defined as habitats within 1.3 m of the surface, the average depth of the photic zone in today's system; Fleenor et al. (preprint).

43   Howe et al. 2014.

44   After Robinson et al. 2016, Appendix C.

45   Cardinale et al. 2012.

46   Cardinale et al. 2012, Hooper et al. 2012.

47   Cardinale et al. 2012, Hooper et al. 2012.

48   See Kukkala and Moilanen 2013 and Margules and Pressey 2000 for background on systematic conservation planning. Groves et al. 2002 lays out these three categories of conservation targets.

49   Cohen and Carlton 1995.

REFERENCES

Abensperg-Traun, M., and G. T. Smith. 1999. How Small Is Too Small for Small Animals? Four Terrestrial Arthropod Species in Different-Sized Remnant Woodlands in Agricultural Western Australia. Biodiversity & Conservation 8:709-726.

Ahearn, D. S., J. H. Viers, J. F. Mount, and R. A. Dahlgren. 2006. Priming the Productivity Pump: Flood Pulse Driven Trends in Suspended Algal Biomass Distribution across a Restored Floodplain. Freshwater Biology 51:1417-1433.

Ahern, J. 2011. From Fail-Safe to Safe-to-Fail: Sustainability and Resilience in the New Urban World. Landscape and Urban Planning 100:341-343.

Alberti, M., and J. M. Marzluff. 2004. Ecological Resilience in Urban Ecosystems: Linking Urban Patterns to Human and Ecological Functions. Urban Ecosystems 7:241-265.

Allen-Diaz, B., R. B. Standiford, and R. D. Jackson. 2007. Oak Woodlands and Forest, Chapter 12. Pages 313-338 in M. Barbour, T. Keeler-Wolf, and A. Schoenherr, editors. Terrestrial Vegetation of California. 3rd Edition. University of California Press, Berkeley, Los Angeles, London.

Amezaga, J., L. Santamaría, and A. J. Green. 2002. Biotic Wetland Connectivity—Supporting a New Approach for Wetland Policy. Acta Oecologica 23:213-222.

Anderson, F. E., B. Bergamaschi, C. Sturtevant, S. Knox, L. Hastings, L. Windham☐Myers, M. Detto, E. L. Hestir, J. Drexler, and R. L. Miller. 2016. Variation of Energy and Carbon Fluxes from a Restored Temperate Freshwater Wetland and Implications for Carbon Market Verification Protocols. Journal of Geophysical Research: Biogeosciences.

Anderson, M., and C. E. Ferree. 2010. Conserving the Stage: Climate Change and the Geophysical Underpinnings of Species Diversity. PLoS ONE 5:1-10.

Atwater, B. F., S. G. Conard, J. N. Dowden, C. W. Hedel, R. L. Macdonald, and W. Savage. 1979. History, Landforms, and Vegetation of the Estuary's Tidal Marshes. Page 493 p. in T. J. Conomos, editor. San Francisco Bay : The Urbanized Estuary : Investigations into the Natural History of San Francisco Bay and Delta with Reference to the Influence of Man : Fifty-Eighth Annual Meeting of the Pacific Division/American Association for the Advancement of Science Held at San Francisco State University, San Francisco, California, June 12-16, 1977. AAAS, Pacific Division, San Francisco, Calif.

Barwis, J. H. 1977. Sedimentology of Some South Carolina Tidal-Creek Point Bars, and a Comparison with Their Fluvial Counterparts. Fluvial Sedimentology Memoir 5:129-160.

Bastian, R., and D. Hammer. 1993. The Use of Constructed Wetlands for Wastewater Treatment and Recycling. Pages 59-68 in G. A. Moshiri, editor. Constructed Wetlands for Water Quality Improvement. Lewis Publishers, Boca Raton, Florida.

Bate, G., A. Whitfield, J. Adams, P. Huizinga, and T. Wooldridge. 2002. The Importance of the River-Estuary Interface (REI) Zone in Estuaries. Water SA 28:271-280.

Bates, M. E., and J. R. Lund. 2013. Delta Subsidence Reversal, Levee Failure, and Aquatic Habitat—a Cautionary Tale. San Francisco Estuary and Watershed Science 11.

Beechie, T. J., D. A. Sear, J. D. Olden, G. R. Pess, J. M. Buffington, H. Moir, P. Roni, and M. M. Pollock. 2010. Process-Based Principles for Restoring River Ecosystems. BioScience 60:209-222.

Beedy, E. 1989. Draft Habitat Suitability Index Model, Tricolored Blackbird (*Agelaius Tricolor*). Prepared by Jones & Stokes Associates for US Bureau of Reclamation, Sacramento, CA.

Beller, E., A. Robinson, R. Grossinger, and L. Grenier. 2015. Landscape Resilience Framework: Operationalizing Ecological Resilience at the Landscape Scale. San Francisco Estuary Institute-Aquatic Science Center (SFEI-ASC), Richmond, CA.

Booth, D. B., J. G. Scholz, T. J. Beechie, and S. C. Ralph. 2016. Integrating Limiting-Factors Analysis with Process-Based Restoration to Improve Recovery of Endangered Salmonids in the Pacific Northwest, USA. Water 8:174.

Boulton, A. J., T. Datry, T. Kasahara, M. Mutz, and J. A. Stanford. 2010. Ecology and Management of the Hyporheic Zone: Stream-Groundwater Interactions of Running Waters and Their Floodplains. Journal of the North American Benthological Society 29:26-40.

Boulton, A. J., S. Findlay, P. Marmonier, E. H. Stanley, and H. M. Valett. 1998. The Functional Significance of the Hyporheic Zone in Streams and Rivers. Annual Review of Ecology and Systematics:59-81.

Boyles, J. G., P. M. Cryan, G. F. McCracken, and T. H. Kunz. 2011. Economic Importance of Bats in Agriculture. Science 332:41-42.

Brett, J. R. 1952. Temperature Tolerance in Young Pacific Salmon, Genus *Oncorhynchus*. Journal of the Fisheries Board of Canada 9:265-323.

Brinson, M. M., R. R. Christian, and L. K. Blum. 1995. Multiple States in the Sea-Level Induced Transition from Terrestrial Forest to Estuary. Estuaries 18:648-659.

Brown, K. J., and G. B. Pasternack. 2005. A Palaeoenvironmental Reconstruction to Aid in the Restoration of Floodplain and Wetland Habitat on an Upper Deltaic Plain, California, USA. Foundation for Environmental Conservation 32:103-116.

Brown, L. R., and D. Michniuk. 2007. Littoral Fish Assemblages of the Alien-Dominated Sacramento-San Joaquin Delta, California, 1980–1983 and 2001–2003. Estuaries and Coasts 30:186-200.

Budd, R., A. O'Geen, K. S. Goh, S. Bondarenko, and J. Gan. 2009. Efficacy of Constructed Wetlands in Pesticide Removal from Tailwaters in the Central Valley, California. Environmental Science & Technology 43:2925-2930.

Burau, J., A. Blake, and R. Perry. 2007. Sacramento/San Joaquin River Delta Regional Salmon Outmigration Study Plan: Developing Understanding for Management and Restoration.

Butte Regional Conservation Plan (BRCP). 2015. Butte Regional Conservation Plan Formal Public Draft BRCP.

Cable Rains, M., G. E. Fogg, T. Harter, R. A. Dahlgren, and R. J. Williamson. 2006. The Role of Perched Aquifers in Hydrological Connectivity and Biogeochemical Processes in Vernal Pool Landscapes, Central Valley, California. Hydrological Processes 20:1157-1175.

California Department of Water Resources, United States Bureau of Reclamation, United States Fish and Wildlife Service, and National Marine Fisheries Service. 2013. Draft Environmental Impact Report / Environmental Impact Statement (DEIR/EIS) for the Bay Delta Conservation Plan. Prepared by by ICF International.

California Department of Water Resources, United States Bureau of Reclamation, United States Fish and Wildlife Service, and National Marine Fisheries Service. 2015. Public Review Partially Recirculated Draft Environmental Impact Report/Supplemental Draft Environmental Impact Statement (RDEIR/SDEIS). Prepared by by ICF International.

California Department of Water Resources (DWR). 2013. Bay Delta Conservation Plan (Public Draft). Prepared by ICF International Sacramento, CA.

Campbell Grant, E. H., W. H. Lowe, and W. F. Fagan. 2007. Living in the Branches: Population Dynamics and Ecological Processes in Dendritic Networks. Ecology Letters 10:165-175.

Cardinale, B. J., J. E. Duffy, A. Gonzalez, D. U. Hooper, C. Perrings, P. Venail, A. Narwani, G. M. Mace, D. Tilman, and D. A. Wardle. 2012. Biodiversity Loss and Its Impact on Humanity. Nature 486:59-67.

Carpenter, S. R., and W. A. Brock. 2004. Spatial Complexity, Resilience, and Policy Diversity: Fishing on Lake-Rich Landscapes. Ecology and Society 9:8.

Cavallo, B., J. Merz, and J. Setka. 2013. Effects of Predator and Flow Manipulation on Chinook Salmon (*Oncorhynchus Tshawytscha*) Survival in an Imperiled Estuary. Environmental Biology of Fishes 96:393-403.

CBEC. 2010. BDCP Effects Analysis: 2D Hydrodynamic Modeling of the Fremont Weir Diversion Structure. Prepared for SAIC and the California Department of Water Resources.

Cecala, K. K., W. H. Lowe, and J. C. Maerz. 2014. Riparian Disturbance Restricts in-Stream Movement of Salamanders. Freshwater Biology 59:2354-2364.

Center, A. S. 2012. The Pulse of the Delta: Linking Science & Management through Regional Monitoring. Aquatic Science Center, Richmond, CA.

Chaplin-Kramer, R., K. Tuxen-Bettman, and C. Kremen. 2011. Value of Wildland Habitat for Supplying Pollination Services to Californian Agriculture. Rangelands 33:33-41.

Chazdon, R. L. 2003. Tropical Forest Recovery: Legacies of Human Impact and Natural Disturbances. Perspectives in Plant Ecology, Evolution and Systematics 6:51-71.

Chelleri, L., and M. Olazabal. 2012. Multidisciplinary Perspectives on Urban Resilience: A Workshop Report. BC3, Basque Centre for Climate Change.

Chelleri, L., J. J. Waters, M. Olazabal, and G. Minucci. 2015. Resilience Trade-Offs: Addressing Multiple Scales and Temporal Aspects of Urban Resilience. Environment and Urbanization 27:181-198.

Christie, M. R., and L. L. Knowles. 2015. Habitat Corridors Facilitate Genetic Resilience Irrespective of Species Dispersal Abilities or Population Sizes. Evolutionary Applications 8:454-463.

Cloern, J. E. 2007. Habitat Connectivity and Ecosystem Productivity: Implications from a Simple Model. The American Naturalist 169:E000.

Cloern, J. E., Robinson, A., Richey A., Grenier, L., Grossinger, R., Boyer, K.E., Burau, J., Canuel, E.A., DeGeorge, J.F., Drexler, J.Z., Enright, C., Howe, E.R., Kneib, R., Mueller-Solger, A., Naiman R.J., Pinckney, J.L., Safran, S.M., Shoellhamer, D., Simenstad, C. . (2016). Primary Production in the

Delta: Then and Now. San Francisco Estuary and Watershed Science, 14(3). jmie_sfews_32799. Retrieved from: http://escholarship.org/uc/item/8fq0n5gx

Cloern, J. E., S. Foster, and A. Kleckner. 2014. Phytoplankton Primary Production in the World's Estuarine-Coastal Ecosystems. Biogeosciences 11:2477-2501.

Cloern, J. E., N. Knowles, L. R. Brown, D. Cayan, M. D. Dettinger, T. L. Morgan, D. H. Schoelhamer, M. T. Stacey, M. van der Wegen, R. W. Wagner, and A. D. Jassby. 2011. Projected Evolution of California's San Francisco Bay-Delta-River System in a Century of Climate Change. PLoS ONE 6.

Cohen, A. N., and J. T. Carlton. 1995. Nonindigenous Aquatic Species in a United States Estuary: A Case Study of the Biological Invasions of the San Francisco Bay and Delta. A Report for the United States Fish and Wildlife Service, Washington D.C. and The National Sea Grant College Program, Connecticut Sea Grant.

Collins, J. N., J. L. Grenier, J. Didonato, G. Geupel, T. Kucera, B. Lidicker, B. Rainey, and S. Rottenborn. 2007. Ecological Connections between Baylands and Uplands: Examples from Marin County.

Collins, J. N., M. Sutula, E. D. Stein, M. Odaya, E. Zhang, and K. Larned. 2006. Comparison of Methods to Map California Riparian Areas. Final Report Prepared for the California Riparian Habitat Joint Venture. SFEI Report No. 522. San Francisco Estuary Institute and Southern California Coastal Water Research Project, Oakland, CA.

Conrad, J. L., E. Holmes, C. Jeffres, L. Takata, N. Ikemiyagi, J. Katz, and T. Sommer. 2016. Application of Passive Integrated Transponder Technology to Juvenile Salmon Habitat Use on an Experimental Agricultural Floodplain. North American Journal of Fisheries Management 36:30-39.

Council, D. S. 2013. The Delta Plan: Ensuring a Reliable Water Supply for California, a Healthy Delta Ecosystem, and a Place of Enduring Value.

Cramer, V. A., R. J. Hobbs, and R. J. Standish. 2008. What's New About Old Fields? Land Abandonment and Ecosystem Assembly. Trends in Ecology & Evolution 23:104-112.

Das, T., M. D. Dettinger, D. R. Cayan, and H. G. Hidalgo. 2011. Potential Increase in Floods in California's Sierra Nevada under Future Climate Projections. Climatic Change 109:71-94.

Davies, Z. G., and A. S. Pullin. 2007. Are Hedgerows Effective Corridors between Fragments of Woodland Habitat? An Evidence-Based Approach. Landscape Ecology 22:333-351.

Delta Independent Science Board. 2016. Draft Workshop Report—Earthquakes and High Water as Levee Hazards in the Sacramento-San Joaquin Delta.

Delta Protection Commission. 2013. Delta Working Lands Program Final Report.

Dettinger, M. D., and D. R. Cayan. 1995. Large-Scale Atmospheric Forcing of Recent Trends toward Early Snowmelt Runoff in California. Journal of Climate 8:606-623.

Dettinger, M. D., D. R. Cayan, M. K. Meyer, and A. E. Jeton. 2004. Simulated Hydrologic Responses to Climate Variations and Change in the Merced, Carson, and American River Basins, Sierra Nevada, California, 1900–2099. Climatic Change 62:283-317.

Deverel, S. J., and D. A. Leighton. 2010. Historic, Recent, and Future Subsidence, Sacramento-San Joaquin Delta, California, USA. San Francisco Estuary and Watershed Science 8.

Dietz, M. F. 2007. Low Impact Development Practices: A Review of Current Research and Recommendations for Future Directions. Water, Air, & Soil Pollution 186:351-363.

Doody, J. P. 2007. Saltmarsh Conservation, Management and Restoration. Springer Science & Business Media.

Drever, C. R., G. Peterson, C. Messier, Y. Bergeron, and M. Flannigan. 2006. Can Forest Management Based on Natural Disturbances Maintain Ecological Resilience? Canadian Journal of Forest Research 36:2285-2299.

Drexler, J. Z. 2011. Peat Formation Processes through the Millennia in Tidal Marshes of the Sacramento–San Joaquin Delta, California, USA. Estuaries and Coasts 34:900-911.

Drexler, J. Z., C. S. de Fontaine, and S. J. Deverel. 2009. The Legacy of Wetland Drainage on the Remaining Peat in the Sacramento-San Joaquin Delta, California, USA. Wetlands 29:372-386.

Dunk, J. R., and R. J. Cooper. 1994. Territory-Size Regulation in Black-Shouldered Kites. The Auk 111:588-595.

Elmore, A. J., S. J. Manning, J. F. Mustard, and J. M. Craine. 2006. Decline in Alkali Meadow Vegetation Cover in California: The Effects of Groundwater Extraction and Drought. Journal of Applied Ecology 43:770-779.

Elmqvist, T., C. Folke, M. Nyström, G. Peterson, J. B. B. Walker, and J. Norberg. 2003. Response Diversity, Ecosystem Change, and Resilience. Frontiers in Ecology 1:488-494.

Elphick, C. S., P. Baicich, K. C. Parsons, M. Fasola, and L. Mugica. 2010. The Future for Research on Waterbirds in Rice Fields. Waterbirds 33:231-243.

Elsholz, C. R. 2010. Riparian Brush Rabbit Habitat Requirements in Caswell Memorial State Park.

Enright, C. 2008. Tidal Slough "Geometry" Filters Estuarine Drivers, Mediates Transport Processes, and Controls Variability of Ecosystem Gradients in CALFED Science Conference proceeding, Sacramento CA.

Enright, C., and S. D. Culberson. 2009. Salinity Trends, Variability, and Control in the Northern Reach of the San Francisco Estuary. San Francisco Estuary and Watershed Science 7.

Enright, C., S. D. Culberson, and J. R. Burau. 2013. Broad Timescale Forcing and Geomorphic Mediation of Tidal Marsh Flow and Temperature Dynamics. Estuaries and Coasts 36:1319-1339.

Ensign, S. H., M. W. Doyle, and M. F. Piehler. 2013. The Effect of Tide on the Hydrology and Morphology of a Freshwater River. Earth Surface Processes and Landforms 38:655-660.

Estep, J. A. 1989. Biology, Movements, and Habitat Relationships of the Swainson's Hawk in the Central Valley of California, 1986-87. State of California, The Resource Agency, Department of Fish and Game, Wildlife Management Division.

Feyrer, F., and M. P. Healey. 2003. Fish Community Structure and Environmental Correlates in the Highly Altered Southern Sacramento-San Joaquin Delta. Environmental Biology of Fishes 66:123-132.

Feyrer, F., T. Sommer, and W. Harrell. 2006. Managing Floodplain Inundation for Native Fish: Production Dynamics of Age-0 Splittail (*Pogonichthys Macrolepidotus*) in California's Yolo Bypass. Hydrobiologia 573:213-226.

Fischer, J., and D. B. Lindenmayer. 2007. Landscape Modification and Habitat Fragmentation: A Synthesis. Global Ecology and Biogeography 16:265-280.

Fischer, J., D. B. Lindenmayer, and A. D. Manning. 2006. Biodiversity, Ecosystem Function, and Resilience: Ten Guiding Principles for Commodity Production Landscapes. Frontiers in Ecology and the Environment 4:80-86.

Fleenor, W., S. Safran, A. Bell, S. Andrews. (preprint). Producing a Pre-Development (ca. 1800) Bathymetric-Topographic Digital Elevation Model of the Sacramento-San Joaquin Delta.

Florsheim, J. L., J. F. Mount, and A. Chin. 2008. Bank Erosion as a Desirable Attribute of Rivers. BioScience 58.

Florsheim, J. L., J. F. Mount, C. Hammersmark, W. E. Fleenor, and G. S. Schladow. 2008. Geomorphic Influence on Flood Hazards in a Lowland Fluvial-Tidal Transitional Area, Central Valley, California. Natural Hazards Review 9:116-124.

Gaines, D. 1977. The Valley Riparian Forests of California: Their Importance to Bird Populations. Pages 57-85 in A. Sands, editor. Riparian Forests in California: Their Ecology and Conservation. Institute of Ecology, University of California, Davis.

Gersberg, R., B. Elkins, S. Lyon, and C. Goldman. 1986. Role of Aquatic Plants in Wastewater Treatment by Artificial Wetlands. Water Research 20:363-368.

Gervais, J. A., D. K. Rosenberg, D. M. Fry, L. Trulio, and K. K. Sturm. 2000. Burrowing Owls and Agricultural Pesticides: Evaluation of Residues and Risks for Three Populations in California, USA. Environmental Toxicology and Chemistry 19:337-343.

Gilroy, J. J., F. A. Edwards, C. A. Medina Uribe, T. Haugaasen, and D. P. Edwards. 2014. Editor's Choice: Surrounding Habitats Mediate the Trade-Off between Land-Sharing and Land-Sparing Agriculture in the Tropics. Journal of Applied Ecology 51:1337-1346.

Goals Project. 2015. The Baylands and Climate Change: What We Can Do. Baylands Ecosystem Habitat Goals Science Update 2015. California State Coastal Conservancy, Oakland, CA.

Grosholz, E., and E. Gallo. 2006. The Influence of Flood Cycle and Fish Predation on Invertebrate Production on a Restored California Floodplain. Hydrobiologia 568:91-109.

Grossinger, R. M., and A. Whipple. 2009. Re-Oaking the Valleys: Bringing Native Trees Back into California's Suburban Landscapes. State of the Estuary Conference, Oakland, CA.

Groves, C. R., D. B. Jensen, L. L. Valutis, K. H. Redford, M. L. Shaffer, J. M. Scott, J. V. Baumgartner, J. V. Higgins, M. W. Beck, and M. G. Anderson. 2002. Planning for Biodiversity Conservation: Putting Conservation Science into Practice. BioScience 52:499-512.

Gunderson, L. H. 2000. Ecological Resilience--in Theory and Application. Annual Review of Ecology and Systematics 31:425-439.

Hall, L. A. 2015. Linked Landscapes: Metapopulation Connectivity of Secretive Wetland Birds. University of California, Berkeley.

Hellmann, J. J., J. E. Byers, B. G. Bierwagen, and J. S. Dukes. 2008. Five Potential Consequences of Climate Change for Invasive Species. Conservation Biology 22:534-543.

Herzog, S. K. 1996. Wintering Swainson's Hawks in California's Sacramento-San Joaquin River Delta. The Condor 98:876-879.

Higgs, E., D. A. Falk, A. Guerrini, M. Hall, J. Harris, R. Jobbs, S. Jackson, J. Rhemtulla, and W. Throop. 2014. The Changing Role of History in Restoration Ecology. Frontiers in Ecology 12:499-506.

Hinsley, S. A., and P. E. Bellamy. 2000. The Influence of Hedge Structure, Management and Landscape Context on the Value of Hedgerows to Birds: A Review. Journal of Environmental Management 60:33-49.

Hobbs, R. J., S. Arico, J. Aronson, J. S. Baron, P. Bridgewater, V. A. Cramer, P. R. Epstein, J. J. Ewel, C. A. Klink, A. E. Lugo, D. Norton, D. Ojima, D. M. Richardson, E. W. Sanderson, F. Valladares, M. Vilà, R. Zamora, and M. Zobel. 2006. Novel Ecosystems: Theoretical and Management Aspects of the New Ecological World Order. Global Ecology and Biogeography 15:1-7.

Hobbs, R. J., E. Higgs, C. M. Hall, P. Bridgewater, F. S. Chapin, E. C. Ellis, J. J. Ewel, L. M. Hallett, J. Harris, and K. B. Hulvey. 2014. Managing the Whole Landscape: Historical, Hybrid, and Novel Ecosystems. Frontiers in Ecology and the Environment 12:557-564.

Hood, W. 2007. Scaling Tidal Channel Geometry with Marsh Island Area: A Tool for Habitat Restoration, Linked to Channel Formation Process. Water Resources Research 43.

Hooper, D. U., E. C. Adair, B. J. Cardinale, J. E. Byrnes, B. A. Hungate, K. L. Matulich, A. Gonzalez, J. E. Duffy, L. Gamfeldt, and M. I. O'Connor. 2012. A Global Synthesis Reveals Biodiversity Loss as a Major Driver of Ecosystem Change. Nature 486:105-108.

Howe, E. 2014. Unraveling Sources of Food Web Support in the Sacramento-San Joaquin Delta's Marsh Ecosystems Using Fatty Acid Biomarkers and Multiple Stable Isotopes. in C. Simenstad and M. Young, editors.

Ingebritsen, S., M. Ikehara, D. Galloway, and D. Jones. 2000. Delta Subsidence in California: The Sinking Heart of the State. 2327-6932, Geological Survey (US).

Ivey, G., B. Dugger, C. Herziger, M. Casazza, and J. Fleskes. 2011. Sandhill Crane Use of Agricultural Lands in the Sacramento–San Joaquin Delta. Final report to Bay-Delta Authority, Sacramento, California, USA.

Jassby, A. D., and J. E. Cloern. 2000. Organic Matter Sources and Rehabilitation of the Sacramento-San Joaquin Delta (California, USA). Aquatic Conservation: Marine and Freshwater Ecosystems 10:323-352.

Jassby, A. D., J. E. Cloern, and B. E. Cole. 2002. Annual Primary Production: Patterns and Mechanisms of Change in a Nutrient-Rich Tidal Ecosystem. Limnology and Oceanography 47:698-712.

Johnston, D. W., and E. P. Odum. 1956. Breeding Bird Populations in Relation to Plant Succession on the Piedmont of Georgia. Ecology 37:50-62.

Jokimäki, J. 1999. Occurrence of Breeding Bird Species in Urban Parks: Effects of Park Structure and Broad-Scale Variables. Urban Ecosystems 3:21-34.

Jones, J. A., F. J. Swanson, B. C. Wemple, and K. U. Snyder. 2000. Effects of Roads on Hydrology, Geomorphology, and Disturbance Patches in Stream Networks. Conservation Biology 14:76-85.

Katz, J., C. Jeffres, L. Conrad, T. Sommer, N. Corline, J. Martinez, S. Brumbaugh, L. Takata, N. Ikemiyagi, and J. Kiernan. 2013. Experimental Agricultural Floodplain Habitat Investigation at Knaggs Ranch on Yolo Bypass, 2012–2013. Sacramento (CA): US Bureau of Reclamation.

Keevil, C., D. Parsons, P. Ashworth, J. Best, S. Sandbach, G. Sambrook Smith, E. Prokocki, A. Nichalas, and C. Simpson. 2013. Flow Structure and Bedform Dynamics around Tidally-Influenced Bars. Presented at: Marine and River Dyne Dynamics, 15 and 16 of April 2013, Bruges, Belgium.

Kelly, J. P., D. Stralberg, K. Etienne, and M. McCaustland. 2008. Landscape Influence on the Quality of Heron and Egret Colony Sites. Wetlands 28:257-275.

Kie, J. G., R. T. Bowyer, M. C. Nicholson, B. B. Boroski, and E. R. Loft. 2002. Landscape Heterogeneity at Differing Scales: Effects on Spatial Distribution of Mule Deer. Ecology 83:530-544.

Kiernan, J. D., P. B. Moyle, and P. K. Crain. 2012. Restoring Native Fish Assemblages to a Regulated California Stream Using the Natural Flow Regime Concept. Ecological Applications 22:1472-1482.

Kirwan, M. L., and A. B. Murray. 2007. A Coupled Geomorphic and Ecological Model of Tidal Marsh Evolution. Proceedings of the National Academy of Sciences 104:6118-6122.

Kjelson, M. A., P. F. Raquel, and F. W. Fisher. 1982. Life History of Fall-Run Juvenile Chinook Salmon, *Oncorhynchus Tshawytscha*, in the Sacramento-San Joaquin Estuary, California.

Kneib, R. T., C. A. Simenstad, M. L. Nobriga, and D. M. Talley. 2008. Tidal Marsh Conceptual Model, Sacramento-San Joaquin Delta Regional Ecosystem Restoration Implementation Plan.

Kondolf, G., and J. Williams. 1999. Flushing Flows: A Review of Concepts Relevant to Clear Creek, California. Report to US Fish and Wildlife Service, Red Bluff, CA.

Kondolf, G. M., A. J. Boulton, S. O'Daniel, G. C. Poole, F. J. Rahel, E. H. Stanley, E. Wohl, A. Bång, J. Carlstrom, C. Cristoni, H. Huber, S. Koljonen, P. Louhi, and K. Nakamura. 2006. Process-Based Ecological River Restoration: Visualizing Three-Dimensional Connectivity and Dynamic Vectors to Recover Lost Linkages. Ecology and Society 11:5.

Kondolf, G. M., and P. R. Wilcock. 1996. The Flushing Flow Problem: Defining and Evaluating Objectives. Water Resources Research 32:2589-2599.

Kukkala, A. S., and A. Moilanen. 2013. Core Concepts of Spatial Prioritisation in Systematic Conservation Planning. Biological Reviews 88:443-464.

Laymon, S. A., M. D. Halterman, and M. Halterman. 1989. A Proposed Habitat Management Plan for Yellow-Billed Cuckoos in California. Forest Service General Technical Report PSW-110:272-277.

Longcore, T., and K. H. Osborne. 2015. Butterflies Are Not Grizzly Bears: Lepidoptera Conservation in Practice. Pages 161-192 in C. J. Daniels, editor. Butterfly Conservation in North America: Efforts to Help Save Our Charismatic Microfauna. Springer Netherlands, Dordrecht.

Luery, M. 2013. Changes Proposed to Calif. Twin Tunnel Project. KCRA, Sacramento, CA.

Lund, J., E. Hanak, W. Fleenor, W. Bennett, and R. Howitt. 2010. Comparing Futures for the Sacramento-San Joaquin Delta. University of California Press.

Lund, J., E. Hanak, W. Fleenor, R. Howitt, J. Mount, and P. Moyle. 2007. Envisioning Futures for the Sacramento-San Joaquin Delta. Public Policy Institute of California.

Luoma, S. N., C. N. Dahm, M. Healey, and J. N. Moore. 2015. Challenges Facing the Sacramento–San Joaquin Delta: Complex, Chaotic, or Simply Cantankerous? San Francisco Estuary and Watershed Science 13.

MacFarlane, R. B., and E. C. Norton. 2002. Physiological Ecology of Juvenile Chinook Salmon (*Oncorhynchus Tshawytscha*) at the Southern End of Their Distribution, the San Francisco Estuary and Gulf of the Farallones, California. Fishery Bulletin 100:244-257.

Malamud-Roam, F., B. Ingram, M. Hughes, and J. Florsheim. 2006. Late Holocene Paleoclimate Records from the San Francisco Bay Estuary and Watershed, California. Quaternary Science Reviews 25:1570–1598.

Malamud-Roam, F., and B. L. Ingram. 2004. Late Holocene $\Delta 13c$ and Pollen Records of Paleosalinity from Tidal Marshes in the San Francisco Bay Estuary, California. Quaternary Research 62:134-145.

Malamud-Roam, F. P., B. L. Ingram, M. Hughes, and J. L. Florsheim. 2006. Holocene Paleoclimate Records from a Large California Estuarine System and Its Watershed Region: Linking Watershed Climate and Bay Conditions. Quaternary Science Reviews 25:1570-1598.

Margules, C. R., and R. L. Pressey. 2000. Systematic Conservation Planning. Nature 405:243-253.

Matella, M. K., and A. M. Merenlender. 2014. Scenarios for Restoring Floodplain Ecology Given Changes to River Flows under Climate Change: Case from the San Joaquin River, California. River Research and Applications 31:280-290.

Mawdsley, J. R., R. O'Malley, and D. S. Ojima. 2009. A Review of Climate-Change Adaptation Strategies for Wildlife Management and Biodiversity Conservation. Conservation Biology 23:1080-1089.

McKinney, M. L. 2006. Urbanization as a Major Cause of Biotic Homogenization. Biological Conservation 127:247-260.

Mclaughlin, B. C., and E. S. Zavaleta. 2012. Predicting Species Responses to Climate Change: Demography and Climate Microrefugia in California Valley Oak (*Quercus Lobata*). Global Change Biology 18:2301-2312.

Meadows, R. 2014. Pop-up Wetlands Help Birds through CA Drought. Frontiers in Ecology and the Environment 12:540-544.

Messick, J. P., and M. G. Hornocker. 1981. Ecology of the Badger in Southwestern Idaho. Wildlife Monographs:3-53.

Miller, R. L., M. Fram, R. Fujii, and G. Wheeler. 2008. Subsidence Reversal in a Re-Established Wetland in the Sacramento-San Joaquin Delta, California, USA. San Francisco Estuary and Watershed Science 6.

Monsen, N. E., J. E. Cloern, and J. R. Burau. 2007. Effects of Flow Diversions on Water and Habitat Quality: Examples from California's Highly Manipulated Sacramento-San Joaquin Delta. San Francisco Estuary and Watershed Science 5.

Morgan-King, T. L., and D. H. Schoellhamer. 2013. Suspended-Sediment Flux and Retention in a Backwater Tidal Slough Complex near the Landward Boundary of an Estuary. Estuaries and Coasts 36:300-318.

Mount, J., and R. Twiss. 2005. Subsidence, Sea Level Rise, and Seismicity in the Sacramento-San Joaquin Delta. San Francisco Estuary and Watershed Science 3.

Moyle, P., W. Bennett, J. Durand, W. Fleenor, B. Bray, E. Hanak, J. Lund, and J. Mount. 2012. Where the Wild Things Aren't: Reconciling the Sacramento-San Joaquin Delta Ecosystem. Public Policy Institute of California.

Moyle, P. B. 2002 Inland Fishes of California. Revised and Expanded. University of California Press, Berkeley, CA.

Moyle, P. B. 2008. The Future of Fish in Response to Large-Scale Change in the San Francisco Estuary, California.in American Fisheries Society Symposium.

Moyle, P. B. 2014. Novel Aquatic Ecosystems: The New Reality for Streams in California and Other Mediterranean Climate Regions. River Research and Applications 30:1335-1344.

Moyle, P. B., B. Herbold, D. E. Stevens, and L. W. Miller. 1992. Life History and Status of Delta Smelt in the Sacramento-San Joaquin Estuary, California. Transactions of the American Fisheries Society 121:67-77.

National Research Council (NRC). 2012. Sea-Level Rise for the Coasts of California, Oregon, and Washinton: Past, Present, and Future. National Research Council, Committee on Sea Level Rise in California, Oregon, and Washington, Washington, D.C.

Nobriga, M. L., and F. Feyrer. 2007. Shallow-Water Piscivore-Prey Dynamics in California's Sacramento-San Joaquin Delta. San Francisco Estuary and Watershed Science 5.

Nobriga, M. L., F. Feyrer, R. D. Baxter, and M. Chotkowski. 2005. Fish Community Ecology in an Altered River Delta: Spatial Patterns in Species Composition, Life History Strategies, and Biomass. Estuaries 28:776-785.

Nyström, M. 2006. Redundancy and Response Diversity of Functional Groups: Implications for the Resilience of Coral Reefs. AMBIO: A Journal of the Human Environment 35:30-35.

Odum, E. 1980. The Status of Three Ecosystem-Level Hypotheses Regarding Salt Marsh Estuaries: Tidal Subsidy, Outwelling and Detritus Based Food Chains. Pages 485-495 in V. Kennecy, editor. Estuarine Perspectives. Academic Press, New York.

Odum, W. E. 1988. Comparative Ecology of Tidal Freshwater and Salt Marshes. Annual Review of Ecology and Systematics 19:147-176.

Odum, W. E. 1990. Internal Processes Influencing the Maintenance of Ecotones: Do They Exist. The ecology and management of aquatic-terrestrial ecotones:91-102.

Olds, A. D., R. M. Connolly, K. A. Pitt, and P. S. Maxwell. 2012. Habitat Connectivity Improves Reserve Performance. Conservation Letters 5:56-63.

Oliver, T. H., T. Brereton, and D. B. Roy. 2013. Population Resilience to an Extreme Drought Is Influenced by Habitat Area and Fragmentation in the Local Landscape. Ecography 35:1-8.

Opperman, J. J., R. Luster, B. A. McKenney, M. Roberts, and A. W. Meadows. 2010. Ecologically Functional Floodplains: Connectivity, Flow Regime and Scale. Journal of the American Water Resources Association 46:211-226.

Peterson, G., C. R. Allen, and C. Holling. 1998. Ecological Resilience, Biodiversity, and Scale. Ecosystems 1:6-18.

Pilliod, D. S., J. L. Welty, and R. Stafford. 2013. Terrestrial Movement Patterns of Western Pond Turtles (*Actinemys Marmorata*) in Central California. Herpetological Conservation and Biology 8:207-221.

Prichard, T. 1979. Agriculture in the Sacramento-San Joaquin Delta. California Agriculture 33:4-5.

Rains, M. C., R. A. Dahlgren, G. E. Fogg, T. Harter, and R. J. Williamson. 2008. Geological Control of Physical and Chemical Hydrology in California Vernal Pools. Wetlands 28:347-362.

Richter, B. D., M. Davis, C. Apse, and C. Konrad. 2012. A Presumptive Standard for Environmental Flow Protection. River Research and Applications 28:1312-1321.

Risk, B. B., P. De Valpine, and S. R. Beissinger. 2011. A Robust Design Formulation of the Incidence Function Model of Metapopulation Dynamics Applied to Two Species of Rails. Ecology 92:462-474.

Robinson, A., A. Richey, J. Cloern, K. Boyer, J. Burau, E. Canuel, J. DeGeorge, J. Drexler, L. Grenier, R. Grossinger, E. Howe, R. Kneib, R. Naiman, A. Mueller-Solger, J. Pinckney, D. Schoellhamer, and C. Simenstad. 2016. Primary Production in the Sacramento-San Joaquin Delta: A Science Strategy to Quantify Change and Identify Future Potential., SFEI-ASC's Resilient Landscapes Program, Richmond, CA.

Roe, J. H., and A. Georges. 2007. Heterogeneous Wetland Complexes, Buffer Zones, and Travel Corridors: Landscape Management for Freshwater Reptiles. Biological Conservation 135:67-76.

Rundle, S., M. Attrill, and A. Arshad. 1998. Seasonality in Macroinvertebrate Community Composition across a Neglected Ecological Boundary, the Freshwater-Estuarine Transition Zone. Aquatic Ecology 32:211-216.

Sabalo, R. 2016. Project Aims to Feed Delta Smelt – 'They're Starving to Death'. The Sacramento Bee, Sacramento, CA.

Safran, S., L. Grenier, and R. Grossinger. 2016. Ecological Implications of Modeled Hydrodynamic Changes in the Upper San Francisco Estuary. SFEI Publication #786, San Francisco Estuary Institute, Richmond, CA.

San Francisco Estuary Institute-Aquatic Science Center (SFEI-ASC). 2014. A Delta Transformed: Ecological Functions, Spatial Metrics, and Landscape Change in the Sacramento-San Joaquin Delta. Richmond, CA.

San Francisco Estuary Partnership (SFEP). 2015. The State of the Estuary 2015. San Francisco Estuary Partnership, Oakland, CA.

Schneider, K. S., G. M. Kondolf, and A. Falzone. 2000. Channel-Floodplain Disconnection on the Stanislaus River: A Hydrologic and Geomorphic Perspective. University of California, Berkeley 94720.

Schoellhamer, D. H., S. A. Wright, and J. Drexler. 2012. A Conceptual Model of Sedimentation in the Sacramento–San Joaquin Delta. San Francisco Estuary and Watershed Science 10.

Searcy, C., and H. Shaffer. 2011. Determining the Migration Distance of a Vagile Vernal Pool Specialist: How Much Land Is Required for Conservation of California Tiger Salamanders. Research and recovery in vernal pool landscapes. Studies from the Herbarium:73-87.

Semlitsch, R. D., and J. R. Bodie. 2003. Biological Criteria for Buffer Zones around Wetlands and Riparian Habitats for Amphibians and Reptiles. Conservation Biology 17:1219-1228.

Sgro, C. M., A. J. Lowe, and A. A. Hoffmann. 2011. Building Evolutionary Resilience for Conserving Biodiversity under Climate Change. Evolutionary Applications 4:326-337.

Siegel, S., C. Toms, D. Gillenwater, and C. Enright. 2010. Suisun Marsh Tidal Marsh and Aquatic Habitats Conceptual Model. Chapter 3: Tidal Marsh. In-Progress Draft. Suisun Marsh Habitat Management, Restoration and Preservation Plan.

Simenstad, C., D. Reed, and M. Ford. 2006. When Is Restoration Not? Incorporating Landscape-Scale Processes to Restore Self-Sustaining Ecosystems in Coastal Wetland Restoration. Ecological Engineering 26:27-39.

Smith, D., S. Harrison, C. Firth, and J. Jordan. 2011. The Early Holocene Sea Level Rise. Quaternary Science Reviews 30:1846-1860.

Smith, D. W., and W. L. Verrill. 1998. Vernal Pool-Soil-Landform Relationships in the Central Valley, California.in C. W. Witham, E. T. Bauder, D. Belk, W. R. Ferren, Jr., and R. Ornduff, editors. Ecology, Conservation, and Management of Vernal Pool Ecosystems: Proceedings from a 1996 Conference. California Native Plant Society, Sacramento, CA.

Solomeshch, A. I., M. G. Barbour, and R. Holland. 2007. Vernal Pools. Terrestrial vegetation of California, 3rd edn. University of California Press, California:394-424.

Sommer, T., R. Baxter, and B. Herbold. 1997. Resilience of Splittail in the Sacramento–San Joaquin Estuary. Transactions of the American Fisheries Society 126:961-976.

Sommer, T. R., W. C. Harrell, A. M. Solger, B. Tom, and W. Kimmerer. 2004. Effects of Flow Variation on Channel and Floodplain Biota and Habitats of the Sacramento River, California, USA. Aquatic Conservation: Marine and Freshwater Ecosystems 14:247-261.

Spautz, H., and N. Nur. 2002. Distribution and Abundance in Relation to Habitat and Landscape Features and Nest Site Characteristics of California Black Rail (*Laterallus Jamaicensis Coturniculus*) in the San Francisco Bay Estuary. Point Reyes Bird Observatory, Sacramento, CA.

Spautz, H., N. Nur, D. Stralberg, and Y. Chan. 2006. Multiple-Scale Habitat Relationships of Tidal-Marsh Breeding Birds in the San Francisco Bay Estuary. Studies in Avian Biology 32:247.

Standish, R. J., R. J. Hobbs, M. M. Mayfield, B. T. Bestelmeyer, K. N. Suding, L. L. Battaglia, V. Eviner, C. V. Hawkes, V. M. Temperton, V. A. Cramer, J. A. Harris, J. L. Funk, and P. A. Thomas. 2014. Resilience in Ecology: Abstraction, Distraction, or Where the Action Is? Biological Conservation 177:43-51.

State Water Resources Control Board. 2010. Development of Flow Criteria for the Sacramento-San Joaquin Delta Ecosystem.

Steiger, J., E. Tabacchi, S. Dufour, D. Corenblit, and J. L. Peiry. 2005. Hydrogeomorphic Processes Affecting Riparian Habitat within Alluvial Channel–Floodplain River Systems: A Review for the Temperate Zone. River Research and Applications 21:719-737.

Stevens, M. L., and E. M. Zelazo. 2015. Fire, Floodplains, and Fish: The Historic Ecology of the Lower Cosumnes River Watershed. Pages 155-187 in P.-L. Yu, editor. Rivers, Fish, and the People: Tradition, Science, and Historical Ecology of Fisheries in the American West. The University of Utah Press, Salt Lake City, Utah.

Stralberg, D., M. Brennan, J. C. Callaway, J. K. Wood, L. M. Schile, D. Jongsomjit, M. Kelly, V. T. Parker, and S. Crooks. 2011. Evaluating Tidal Marsh Sustainability in the Face of Sea-Level Rise: A Hybrid Modeling Approach Applied to San Francisco Bay. PLoS ONE 6:e27388.

Sullivan, B. L., J. L. Aycrigg, J. H. Barry, R. E. Bonney, N. Bruns, C. B. Cooper, T. Damoulas, A. A. Dhondt, T. Dietterich, and A. Farnsworth. 2014. The eBird Enterprise: An Integrated Approach to Development and Application of Citizen Science. Biological Conservation 169:31-40.

Swanson, C. 2015. Ecological Processes – Flood Events Indicators Technical Appendix, State of the San Francisco Estuary 2015.

Swanson, C., J. Cain, J. Opperman, and M. Tompkins. 2010. Exhibit TBI-3 before the State Water Resoruces Control Board Written Testimony Regarding Flow Criteria for the Delta Necessary to Protect Public Trust Resources: Delta Inflows.

Swanson, C., R. Turid, P. S. Young, and J. C. Joseph, Jr. 2000. Comparative Environmental Tolerances of Threatened Delta Smelt (*Hypomesus Transpacificus*) and Introduced Wakasagi (*H. Nipponensis*) in an Altered California Estuary. Oecologia 123:384-390.

Teal, J. M. 1962. Energy Flow in the Salt Marsh Ecosystem of Georgia. Ecology:614-624.

Thrush, S., J. Halliday, J. Hewitt, and A. Lohrer. 2008. The Effects of Habitat Loss, Fragmentation, and Community Homogenization on Resilience in Estuaries. Ecological Applications 18:12-21.

Toft, J. D., C. A. Simenstad, J. R. Cordell, and L. F. Grimaldo. 2003. The Effects of Introduced Water Hyacinth on Habitat Structure, Invertebrate Assemblages, and Fish Diets. Estuaries 26:746-758.

Tompkins, E. L., and W. N. Adger. 2004. Does Adaptive Management of Natural Resources Enhance Resilience to Climate Change? Ecology and Society 9:10.

Tremblay, M. A., and C. C. St Clair. 2009. Factors Affecting the Permeability of Transportation and Riparian Corridors to the Movements of Songbirds in an Urban Landscape. Journal of Applied Ecology 46:1314-1322.

Tsao, D. C., R. E. Melcer Jr, and M. Bradbury. 2015. Distribution and Habitat Associations of California Black Rail (*Laterallus Jamaicensis Coturniculus*) in the Sacramento–San Joaquin Delta. San Francisco Estuary and Watershed Science 13.

Tzoulas, K., K. Korpela, S. Venn, V. Yli-Pelkonen, A. Kaźmierczak, J. Niemela, and P. James. 2007. Promoting Ecosystem and Human Health in Urban Areas Using Green Infrastructure: A Literature Review. Landscape and Urban Planning 81:167-178.

U.S. Fish Wildlife Service. 2008. Formal Endangered Species Act Consultation on the Proposed Coordinated Operations of the Central Valley Project (CVP) and State Water Project (SWP).in U.S. Department of the Interior, editor. Memorandum from Regional Director, Fish and Wildlife Service, Region 8, Sacramento, California to Operation Manager, Bureau of Reclamation, Central Valley Operations Office, Sacramento, California, Sacramento, CA.

Van Groenigen, J., E. Burns, J. Eadie, W. Horwath, and C. Van Kessel. 2003. Effects of Foraging Waterfowl in Winter Flooded Rice Fields on Weed Stress and Residue Decomposition. Agriculture, Ecosystems & Environment 95:289-296.

Vogel, D. 2013. Evaluation of Fish Entrainment in 12 Unscreened Sacramento River Diversions. Natural Resource Scientists, Inc.

West, J. M., and J. B. Zedler. 2000. Marsh-Creek Connectivity: Fish Use of a Tidal Salt Marsh in Southern California. Estuaries 23:699-710.

Weston, D. P., A. M. Asbell, S. A. Lesmeister, S. J. Teh, and M. J. Lydy. 2014. Urban and Agricultural Pesticide Inputs to a Critical Habitat for the Threatened Delta Smelt (*Hypomesus Transpacificus*). Environmental Toxicology and Chemistry 33:920-929.

Whipple A, Grossinger RM, Rankin D, Stanford B, and A. R. 2012. Sacramento-San Joaquin Delta Historical Ecology Investigation: Exploring Pattern and Process. San Francisco Estuary Institute-Aquatic Science Center, Richmond, CA.

Whipple, A. A., R. M. Grossinger, and F. W. Davis. 2011. Shifting Baselines in a California Oak Savanna: Nineteenth Century Data to Inform Restoration Scenarios. Restoration Ecology 19:88-101.

Wiens, J. A. 2002. Riverine Landscapes: Taking Landscape Ecology into the Water. Freshwater Biology 47:501-515.

Williams, P. B., E. Andrews, J. J. Opperman, S. Bozkurt, and P. B. Moyle. 2009. Quantifying Activated Floodplains on a Lowland Regulated River: Its Application to Floodplain Restoration in the Sacramento Valley. San Francisco Estuary and Watershed Science 7.

Williams, P. B., and M. K. Orr. 2002. Physical Evolution of Restored Breached Levee Salt Marshes in the San Francisco Bay Estuary. Restoration Ecology 10:527-542.

Wohl, E., P. L. Angermeier, B. Bledsoe, G. M. Kondolf, L. MacDonnell, D. M. Merritt, M. A. Palmer, N. L. Poff, and D. Tarboton. 2005. River Restoration. Water Resources Research 41.

Wu, J. 2014. Urban Ecology and Sustainability: The State-of-the-Science and Future Directions. Landscape and Urban Planning 125:209-221.

Yarnell, S. M., G. E. Petts, J. C. Schmidt, A. A. Whipple, E. E. Beller, C. N. Dahm, P. Goodwin, and J. H. Viers. 2015. Functional Flows in Modified Riverscapes: Hydrographs, Habitats and Opportunities. BioScience 65:963-972.

Yoshiyama, R. M., E. R. Gerstung, F. W. Fisher, and P. B. Moyle. 2001. Historical and Present Distribution of Chinook Salmon in the Central Valley Drainage of California. Contributions to the Biology of Central Valley Salmonids, Fish Bulletin 179:71-176.

Zeiner, D., W. F. Laudenslayer, and K. E. Mayer. 1990. California's Wildlife. Birds. State of California, Resources Agency, Department of Fish and Game.

Zhang, M., and R. Goodhue. 2010. Agricultural Pesticide Best Management Practices Report. A Final Report for the Central Valley Regional Water Quality Control Board.